Genes and DNA

Genes and DNA

A BEGINNER'S GUIDE TO GENETICS

AND ITS APPLICATIONS

CHARLOTTE K. OMOTO AND PAUL F. LURQUIN

COLUMBIA UNIVERSITY PRESS   NEW YORK

Columbia University Press
Publishers Since 1893
New York   Chichester, West Sussex
Copyright © 2004 Columbia University Press
All rights reserved

Library of Congress Cataloging-in-Publication Data
Omoto, Charlotte K.
    Genes and DNA : a beginner's guide to genetics and its applications /
Charlotte K. Omoto and Paul F. Lurquin.
        p.   cm.
    Includes bibliographical references and index.
    ISBN 0-231-13012-0 (cloth : alk. paper)   ISBN 0-231-13013-9
(pbk. : alk. paper)
        1. Genetics.   2. Molecular genetics.   I. Lurquin, Paul F.   II. Title.

QH430.O47 2004
576.5—dc22
        2003062584

∞

Columbia University Press books are printed on permanent
and durable acid-free paper.
Printed in the United States of America
c 10 9 8 7 6 5 4 3 2 1
p 10 9 8 7 6 5 4 3 2 1

IN MEMORY OF OUR MOTHERS

Contents

Acknowledgments

WE APPRECIATE THE HELP of our colleagues Kirstin Malm, Valerie Lynch-Holm, and Norah McCabe in making the images for some of the figures. We appreciate the suggestion from Noreen Warren for testing corn products for genetic modification and for providing more information about Thalidomide. We thank Andris Kleinhofs for help with and images used in box 9.2. Mark Michelsen improved many of the figures to make them suitable for publication. We gratefully thank our colleagues Paul Verrell and Martin Morgan for reading and critiquing a draft of our manuscript. Thanks also to our students who did the same thing in addition to, of course, taking exams on the subject material of this book. Thanks also to three anonymous reviewers whose contributions enhanced our manuscript. Robin Smith, senior editor for the sciences at Columbia University Press, encouraged us at every step of the way. Last but not least, we appreciate the partial financial support from the Honors College and the College of Sciences of Washington State University.

Contributors

DAVID HANSEN
Minnesota Science, University of Minnesota

ANDRIS KLEINHOFS
Department of Crop and Soil Sciences, Washington State
University

VALERIE LYNCH-HOLM
EM Center, Washington State University

KIRSTIN MALM
School of Molecular Biosciences, Washington State University

NORAH MCCABE
School of Molecular Biosciences, Washington State University

NOREEN WARREN
Madison Area Technical College

Preface: Why Is Genetics Important?

PERHAPS YOU'VE HEARD THE MEDIA use the terms "cloning," "genetically modified organisms," "DNA fingerprinting," and "genetic testing." But have you ever discovered what these terms really mean? Our goal is to help you to become more familiar with words like these. We will provide you with clear and straightforward genetic principles that are relevant to your everyday life and help you understand the many applications of modern genetic techniques.

Genetics is one of the greatest adventures in science. This book will help you explore everything from the foundations of genetics, a little over a century ago, to modern genetic applications, including the genetic engineering of plant products that you probably eat on a regular basis. You will learn about medical, legal, and ethical aspects of genetics, as well as the impact of genetics on our society; this impact is mind-boggling. For example, as little as fifteen years ago, it was unthinkable that deoxyribonucleic acid (DNA) would play any role in the prosecution and conviction of criminals. The use of DNA testing is now routine, and many people suggest that DNA testing should be made available for cases that came to trial before this technology was available. Likewise, paternity suits often ended up in mistrials because incontrovertible evidence could not be obtained when only simple blood tests were available. But now no one would consider paternity suits without DNA evidence.

We are discovering many genetic diseases and finding out more about them. The first catalog of human genetic diseases listed fewer than 1,500. Now this catalog is updated almost daily as an Internet resource, part of Online Mendelian Inheritance in Man, with almost 8,000 genetic diseases as of this writing. Every state in the United States, and most of the developed world, now tests newborn babies for genetic diseases, and the number of diseases that are tested for is increasing.

A few years ago, genetically modified corn did not exist. Now, you might eat it for breakfast. As you may know, the existence of genetically modified crops on the shelves of our supermarkets has contributed to street riots in Seattle and New York City. The few examples listed above already demonstrate that the science of genetics is important for society. It is easy to form an opinion about modern genetics without really knowing all that much about it. But we need to be well informed to make good decisions about these important issues. Our ultimate purpose in writing this book is to help you make informed decisions about genetics.

To achieve this goal, you will learn how genes determine the characteristics of all life forms and how these genes are passed from parents to progeny. Next you will find out how genetic analysis was able to proceed before we even knew that DNA exists. You will also examine different types of changes in our genetic material and how these changes affect how we look, act, and feel. You will look at the impact of our environment on our genetic material. Then you will learn how our understanding properties of DNA has led to genetic engineering and its medical applications. You will be introduced to the genetics of populations, in which it is no longer single individuals, but rather large collections of individuals, that are studied. This branch of genetics is the key to the study of species conservation. Population genetics also leads directly to the concepts upon which evolution by natural selection is based. Finally, you will discover the role of genetics and the environment in determining traits. This area of genetics is important in agriculture as well as medicine.

We hope you enjoy learning about genetics with our book. But first and foremost we hope that after reading it, you will understand genetics and will be able to confidently make decisions regarding genetics and genetic technology that affect your life.

Genes and DNA

CHAPTER 1

What Are Genes?

YOU MAY ALREADY KNOW THAT GENES are made of DNA (short for *deoxyribonucleic acid*). More interesting than knowing this is understanding *how* we know that DNA is the basis for heredity and understanding the importance of the structure of DNA for inheritance. You will see in this chapter that DNA and its structure are the keys to understanding inheritance.

DNA

DNA has a fascinating history. The Swiss scientist Friedrich Miescher discovered DNA near the end of the nineteenth century. Miescher never knew that the substance he had isolated from sperm and pus (yes, pus!) would turn out to be so critical to the understanding of life. He died several decades before the function of DNA and its famous double-helical structure were uncovered. After Miescher, other scientists tried to identify the chemical composition of sperm, reasoning that sperm must carry the genetic material to the next generation. These scientists also reasoned that sperm cells have very little excess cellular material other than the hereditary material found in the sperm head. In fact, DNA constitutes over 60 percent of the sperm head; the remainder is mostly protein.

For a long time after Miescher's discovery, DNA was thought to be a simple molecule, consisting of nucleotides strung together like beads on a string. Each nucleotide is composed of a sugar (deoxyribose) chemically linked to phosphorus atoms and one of four different nitrogenous bases (so called because they contain a significant number of nitrogen atoms). The nitrogenous bases are adenine, guanine, cytosine, and thymine. These four bases are abbreviated as A, G, C, and T. Nothing known about the DNA molecule suggested that it could play any role in heredity. The structure of DNA seemed much too simple to account for the many already known hereditary traits. But then scientists found that the building blocks of DNA—the nucleotides—were repeated hundreds of times in the DNA molecule. As techniques to isolate DNA from living cells improved, the number of nucleotides in a DNA molecule was found to be in the thousands, and then in the hundreds of thousands. Scientists had discovered that DNA is a polymer, much like many plastics such as polyethylene and polypropylene, except that DNA is a very long polymer with millions of nucleotides, As, Gs, Cs, and Ts.

Yet nobody knew what living cells did with this polymer, nor did anybody know the structure of the DNA molecule. In fact, some scientists believed that only animals and bacterial cells possessed DNA and that plants were devoid of it. Since plants, as well as animals and bacteria, all had well-defined genetic characteristics (for example, flower color for plants, shape for animals, and pathogenicity for bacteria), DNA could not be the genetic material, so the logic went. We now know that plants do contain DNA, and that the failure to isolate it from them was due to the use of crude techniques. In fact, for geneticists, plants, animals, and bacteria are largely similar in spite of their great diversity. This is because their hereditary properties are all based on the existence of one substance: DNA.

DNA Can Be Specifically Stained and Observed in Cells

An important step in the development of ideas about the chemical nature of the genetic material was the ability to stain DNA. In the 1920s, German biochemist Robert Feulgen developed a way to specifically stain DNA. He then used this method to stain DNA in living tissue. The Feulgen reaction, as it is now called, specifically colors DNA purple. The stained cells can then be viewed under the microscope. Feulgen used this technique on all kinds of tissues from animals, plants, and protozoa. Under the microscope, the purple DNA stain was found in a central compartment of all these cells. The compartment is given the name nucleus, plural nuclei. Feulgen found that the nuclei of all of these cells, including the nuclei of plant cells, became stained. This definitively proved that plants had DNA and that the DNA of cells is located in the nucleus.

With the Feulgen stain, scientists had a tool to measure the amount of DNA present in cells. In 1950, in a paper entitled "Constancy of DNA in Plant Nuclei," Hewson Swift at the University of Chicago showed that all cells from different parts of a corn plant had a constant amount of DNA. Furthermore, the amount of DNA in pollen was half that found in, for example, the leaf and root cells. He found that rapidly dividing cells in the root tip and other cells prior to cell division had twice as much as DNA. These are what one would expect of the genetic material (see chapter 2). If DNA was the genetic material, its amount should be constant in all the cells of the organism regardless of the size of the cell.

An even more interesting observation was made using the Feulgen stain: DNA changes shape as cells divide. Most of the time, the Feulgen stain showed an amorphous purple sphere in the nucleus, without any substructure. But just before cells divide, the DNA becomes condensed into sausage-looking structures called chromosomes. It was found that the number and shape of these condensed chromosomes was the same in different body cells of the same organism. Furthermore, one could see that in sperm cells the number of chromosomes was halved. We would expect that the amount of hereditary material in the gametes, sperm or pollen and egg or ova, would be half that found in the nonreproductive cells of the organism.

DNA Determines Genetic Properties in Bacteria

That DNA is indeed the genetic material was demonstrated in bacteria in 1944. A team led by the Canadian Oswald Avery at Rockefeller University in New York made this landmark discovery. Their biological material was the bacterium *Streptococcus pneumoniae*, which, as its name indicates, causes pneumonia. Avery's laboratory possessed two strains of these bacteria. One strain infects mice with pneumonia (the "virulent" strain), and the other strain does not (the "avirulent" strain). The two strains look different when growing in a petri dish: The virulent strain grows as a smooth, slimy, large collection of cells known as a colony. The avirulent strain produced small, well-defined, rough-looking colonies (figure 1.1). Thus the bacterial strains are distinguishable by two characteristics: physical appearance and the strain's ability or inability to cause pneumonia. These characteristics are hereditary. Avery knew this because he grew the two bacterial strains for a long time, during which the cells divided many times, but this produced no change in the bacteria's ability to cause pneumonia or the bacteria's physical appearance.

Avery did an experiment that suggested that hereditary properties like virulence and appearance could be exchanged between cells. He knew, as we do, that heating bacterial cells kills them. Indeed, when the smooth, virulent bacteria are heated, they are killed and no longer infect mice with pneumonia. But when one mixes these heat-killed virulent bacteria with live, avirulent rough bacteria, one finds that this mixture can kill mice. Avery reasoned that the genetic material from the heat-killed virulent bacteria changed the properties of the rough strain. In this case, he reasoned, DNA purified from the smooth strain would change the characteristics of the rough strain, but only if the DNA could get into the rough, avirulent bacteria. Avery's research team purified DNA from the smooth, slimy strain and added the purified DNA to the rough cells. They observed that a few smooth and slimy cells appeared in the culture of rough cells exposed to the DNA! This "transformation," as they called it, was stable over many cell divisions. Avery and his team interpreted these observations to mean that DNA was the genetic material.

However, to make sure that the DNA did not contain contaminants that could have been the true genetic material, they did further experiments with a variety of enzymes. Suffice it to say that enzymes are proteins that catalyze all sorts of reactions. Biologists commonly give names to these enzymes by putting the suffix "-ase" after the name of the substances they act upon. For example, deoxyribonuclease is an enzyme that destroys DNA by cutting it into its nucleotide building blocks. Similarly, a protease is an enzyme that destroys proteins. When Avery and his coworkers added deoxyribonuclease to their purified DNA and then added this mixture to rough cells, no smooth, slimy cells were recovered. This meant that destroying DNA destroyed the transforming activity. On the other hand, adding proteases to the DNA had no effect on its transforming activity. If contaminating proteins in their DNA samples had been responsible for the transforming activity,

Figure 1.1 Virulent and Avirulent *Streptococcus Pneumoniae* Colonies on an Agar Plate. The large, slimy, smooth colonies on the left are virulent, and the small, distinct colonies on the right are avirulent. Plate and photograph courtesy of Kirstin Malm.

the addition of proteases should have destroyed this activity. It did not. This means that proteins were not responsible for the observed transformation. There it was: the genetic material—the genes—of *Streptococcus pneumoniae* was not made of proteins, but of DNA.

Avery's work had been done under carefully controlled conditions, and his conclusions were straightforward. Yet nobody at the time believed him! Why was that so? It turns out that as recently as the late 1940s, scientists were convinced that protein, not DNA, was the genetic material. Why? Proteins seemed like a much more likely candidate to be the genetic material. Proteins existed in innumerable varieties that differed enormously among living species. For this reason, proteins were thought to constitute the true genetic material. Another factor is that in 1944 World War II was still raging, and Avery's discovery must have been seen as of little consequence when people were dying on the battlefields and in bombed-out cities. Finally, some people thought that DNA was possibly the genetic material of some rare bacterial species, but certainly it could not be responsible for the hereditary properties of higher life forms, such as animals and plants. Of course, the skeptics were dead wrong, as we know. Such an important discovery should have earned its authors a Nobel Prize. However, Avery was sixty-seven at the time he made this important discovery, and he died eleven years later. Recognition of important discoveries often takes decades. Since the Nobel Prize is not granted posthumously, Avery was unable to be recognized for his important work.

We know today that DNA is an almost universal genetic material, and that genes present in simple viruses, bacteria, plants, and animals are all made of DNA. Amazingly, some viruses are made of a chemical very similar to DNA, ribonucleic acid (RNA), where the base thymine (T) is replaced by uracil (U) and where the sugar is ribose, not deoxyribose.

DNA Is a Double Helix

By the late 1940s biochemists knew that DNA was a very long polymer made up of millions of nucleotides. Each nucleotide contains one of the four nitrogenous bases (A, T, G, or C) linked to a deoxyribose unit, in turn linked to a chemical group containing a phosphorus atom. DNA is held together by bonds between the phosphate and the deoxyribose units. Therefore, one speaks of the DNA's "sugar-phosphate backbone" (figure 1.2.A). In those years, it was also known that in all DNA samples isolated from widely different species (human, yeast, and bacteria, for example), the amount of adenine (A) was always equal to the amount of thymine (T). Similarly, guanine (G) was always equal to cytidine (C). Nobody knew how to explain this, but the observation suggested some regularity in the DNA molecule.

A breakthrough occurred when Rosalind Franklin, a researcher at King's College in London, England, succeeded in crystallizing DNA in the early 1950s. Crystals are formed when identical molecules are packed in a very organized fashion. This is rather simple to do for a small molecule. Perhaps you have made sugar crystals by putting a string into a solution saturated with sugar. This happens because the rough structure of the string initiates crystal formation. Once some sugar molecules attach to the string, other sugar molecules can fit in like bricks in a wall. Because DNA is such a large molecule, it does not form crystals readily. Why was it so important to obtain DNA crystals? One can take advantage of the very regular arrangement of the same molecules in a crystal to determine their structure. There existed at the time a well-established technique used to determine the arrangement of atoms inside a crystal. This technique is called X-ray crystallography,

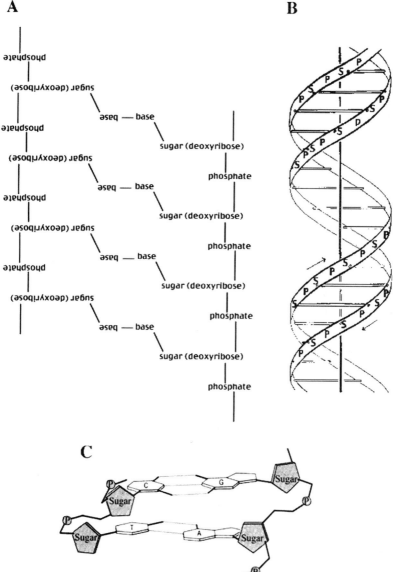

A

phosphate

sugar (deoxyribose)

base — base

phosphate

sugar (deoxyribose)

base — base

phosphate

sugar (deoxyribose)

base — base

phosphate

sugar (deoxyribose)

base — base

sugar (deoxyribose)

phosphate

sugar (deoxyribose)

phosphate

sugar (deoxyribose)

phosphate

sugar (deoxyribose)

phosphate

B

C

Figure 1.2 Diagrams of DNA. A. A flat diagram shows two strands of DNA each with four bases. Note that each strand is held together by a "sugar-phosphate backbone." The two strands run in opposite directions, thus the left-hand strand is shown upside down. The strands are held together by weak bonds, called hydrogen bonds, between the bases. B. A diagram of the double-helix structure of DNA. Note the sugar-phosphate backbone in opposite directions shown by the arrows. The rungs of the ladder represent the bases held together by the weak hydrogen bonds. C. A detailed diagram of two bases from two opposing strands of DNA. The phosphates are shown as shaded circles and sugars as shaded pentagons. The dotted lines connecting the bases are weak hydrogen bonds.

and, as its name indicates, it consists of illuminating a target crystal with X rays. The regular arrangement of atoms in a crystal deflects X rays and forms spots in concentric rings on a photographic film. The more organized the structure, the more spots are formed farther out in the ring. By noting the location and intensity of these spots, one can then determine the relative positions of the atoms in the crystal and determine the three-dimensional structure of the crystallized molecules. Thus, Rosalind Franklin obtained the first high-quality X-ray data for a DNA crystal (figure 1.3).

At this time, James Watson, a young American postdoctoral scientist, and Francis Crick, an English physicist working on his Ph.D. dissertation, were both at Cambridge University. These two struck up a collaboration to solve the problem of the structure of DNA. Neither of them had done any previous work with DNA. They were thus novices, although Crick knew the theory of X-ray crystallography very well. Indeed, Watson and Crick never did a single experiment to solve the structure of DNA. All the experimental results had been obtained by Rosalind Franklin and later repeated by her boss, Maurice Wilkins. One day, while he was visiting the King's College researchers, Watson saw a photograph of DNA made by Franklin. The arrangement of the spots radiating out in an X shape immediately suggested to him that DNA must be a helical molecule. Back in Cambridge, Watson convinced Crick of that interpretation, and model building started. After a few days of trial and error, they had a helical molecule that was also consistent with Franklin's X-ray crystallography data. DNA was a double helix in which the sugar-phosphate backbones are on the outside, while the bases are on the inside of the molecule (figure 1.2.B). This structure was held together by weak bonds between an A and a T, indicated by the dotted lines in the figure, and similarly G was held to a C. This pairing of A with T and G with C is called "complementary base pairing." This explained why the number of As was always equal to the number of Ts, and Gs always equal to Cs. The discovery of the double-helical structure of DNA was published in a short report in 1953, and Watson, Crick, and Wilkins received the Nobel Prize in 1962.

A very sad aspect to this story is that Rosalind Franklin did not receive recognition at the time for her major contribution. She died in 1958 at the young age of thirty-seven of ovarian cancer, unable to share the honors unquestionably due to her too.

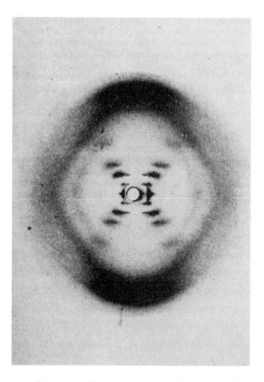

Figure 1.3 Photograph of the X-ray diffraction pattern of DNA. Produced by Rosalind Franklin.

Transfer of Genetic Information to Progeny

In their short article, Watson and Crick announced cryptically that "it has not escaped our notice that the specific [base] pairing we have postulated immediately suggests a possible copying mechanism for the genetic material." What did they mean by that? First, since DNA is the genetic material, it contains all the instructions necessary to "tell" a living cell what it is supposed to do. Next, as cells divide, as for example in the case of bacteria or human cells, the progeny of a cell must contain the same genetic instructions as the original cell. Thus, there must exist a mechanism that copies DNA faithfully, to ensure that progeny cells contain the same genetic material as the maternal cell. When Watson and Crick envisioned a copying mechanism for DNA, none of the details of this mechanism was known. However, simply by looking at the double-helical structure of DNA, they understood the basics of the replication process. To understand how they arrived at this interpretation, we must go back to the complementary base pairing that holds the DNA double helix together: an A always faces a T and a G always faces a C.

If DNA is the genetic material, cells must contain the necessary machinery (like enzymes) that can "read" the base sequence in DNA. With the structure of DNA depicted in figure 1.2, we realize that if the two DNA strands become separated, each strand has the information to specify the order of A, T, G, and C in the other strand. What is the result of this "reading" process? Figure 1.2.C shows that because of the A-T and G-C complementary base-pair arrangement, whenever an A is read in one of the strands of the original DNA molecule, the cellular machinery must incorporate a T in the newly growing opposite strand. But, since the A in the original molecule faced a T, the T in that strand must be read in such a way that the cellular machinery incorporates an A in the other growing strand. The result is that an original A-T pair is now copied into two A-T pairs, each new one present in the two replicated DNA molecules. This copying mechanism occurs at the level of each individual base pair, ensuring that the two resultant double-helical molecules are identical to the original DNA double helix. We say that the two DNA strands are used as templates for the synthesis of two complementary, new DNA strands.

We now know that this is how DNA replicates. The machinery that performs DNA replication is very complex and involves dozens of proteins. One key enzyme in the process of DNA replication is called DNA polymerase. This enzyme "reads" the bases present in the template strands and incorporates the complementary bases into the growing new strands. DNA replication is extremely accurate, but it is not absolutely perfect. Mistakes made by DNA polymerase result in the incorporation of a "wrong" base (like putting in a G opposite an A instead of a T), and these errors are one of the causes of spontaneous mutations, the ultimate source of genetic variation.

DNA Can Be Replicated in the Test Tube
Geneticists now have a good understanding of the many ingredients that are necessary for DNA replication. We need to have the enzyme DNA polymerase to do the job, as well as the building blocks of DNA, nucleotides. In the natural process of DNA replication, the two strands of DNA are separated and the enzyme DNA polymerase binds

to short double-stranded stretches positioned next to the single-stranded DNA to be copied.

Today, it is possible to make significant amounts of DNA in the test tube, using a method that partially imitates the mechanism used by living cells. This method is called the polymerase chain reaction (PCR) and was invented in 1986 by Kary Mullis. Mullis won the Nobel Prize in 1993 and candidly confesses that he came up with the idea on a surfing trip, while high on drugs.

We have seen that each strand of a double-stranded DNA is used as template for DNA replication. DNA normally is held together as a double-stranded molecule by weak bonds between the complementary bases (figure 1.2). It turns out that we can separate the two strands of double-stranded DNA simply by heating up a DNA solution close to the boiling point. At high temperature, the weak bonds that link the two strands together are broken, and the two strands of DNA separate. If the solution is cooled, the double-stranded DNA can reform. As in the cell, DNA polymerase in the test tube needs to have short stretches of double-stranded structure to copy the single stranded regions. By 1986, chemists could make short pieces of single-stranded DNA of a predetermined nucleotide sequence. If we know the base sequence of a piece of DNA, it is possible to synthesize a short piece of DNA whose base sequence is complementary. When the solution is cooled, short strands of DNA can more easily find their complementary strand than long strands. To reform the original long double-stranded DNA, the solution must be cooled slowly. Thus if the solution is fast-cooled, the process of short pieces making a partial double-stranded structure wins out over the long, complementary DNA sequences coming together (figure 1.4.C).

The short pieces of DNA that form the short double-stranded regions are known as "primers." In order to copy both strands of the double-stranded DNA, each strand must have a primer. The relative position of the complementary regions to these primers determines the size of the DNA piece that will be made. As we just saw, these primers can be made in the lab, and they will form weak bonds with single-stranded DNA with a complementary sequence. Therefore, by adding DNA polymerase together with the building blocks of DNA (nucleotides, A, T, G, and C), one can copy DNA in the test tube (figure 1.4). The problem, though, is that one can copy DNA only once

A

Two short regions of a natural double-stranded section of DNA separated by 350 bases shown as dots.

. . T-G-G-T-A-C-T-T-G-A-G-A-C-C-G-G-G-G-C . . . . C-T-C-C-A-G-C-G-G-G-T-C-C-T-T-G-A-T-C-C-A-T-G-T-T-T-A . .
. . | | | | | | | | | | | | | | | | | | | | | | | | | | | | | | | | | | | | | | | | | | | | | | | | | |
. . A-C-C-A-T-G-A-A-C-T-C-T-G-G-C-C-C-C-G . . . . G-A-G-G-T-C-G-C-C-C-A-G-G-A-A-C-T-A-G-G-T-A-C-A-A-A-T .

After heating

. . T-G-G-T-A-C-T-T-G-A-G-A-C-C-G-G-G-G-C . . . . C-T-C-C-A-G-C-G-G-G-T-C-C-T-T-G-A-T-C-C-A-T-G-T-T-T-A . .

. . A-C-C-A-T-G-A-A-C-T-C-T-G-G-C-C-C-C-G . . . . A-G-G-T-C-G-C-C-C-A-G-G-A-A-C-T-A-G-G-T-A-C-A-A-A-T . .

After adding a primer and fast – cooling

. . T-G-G-T-A-C-T-T-G-A-G-A-C-C-G-G-G-G-C . . . . C-T-C-C-A-G-C-G-G-G-T-C-C-T-T-G-A-T-C-C-A-T-G-T-T-T-A . .
                                                        | | | | | | | | | | | | | | | | | | | |
                                                        G-A-G-G-T-C-G-C-C-C-A-G-G-A-A-C-T-A-G-G-T

T-G-G-T-A-C-T-T-G-A-G-A-C-C-G-G-G-G-C
| | | | | | | | | | | | | | | | | | |
. . A-C-C-A-T-G-A-A-C-T-C-T-G-G-C-C-C-C-G . . . . A-G-G-T-C-G-C-C-C-A-G-G-A-A-C-T-A-G-G-T-A-C-A-A-A-T . .

B

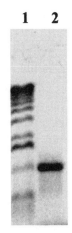

1   2

Figure 1.4 Polymerase Chain Reaction. A. Beginning of a PCR reaction. The top frame depicts a short region of a much longer double-stranded DNA with middle piece of 350 bases represented by dots. When heated, weak bonds between the two strands are broken and the strands separate as shown in the middle frame. When the solution is fast-cooled, primers in the solution bind to the complementary sequences as shown in the bottom frame. Because the primers in this example are 19 and 21 bases long and are separated by 350, this set of primers will make a DNA product that is 390 base pairs long. B. A photograph of the result from a PCR reaction. The DNA runs in a gel from the top toward the bottom during the application of an electric current. 1: DNA size markers with the largest DNA pieces towards the top. 2: PCR product. C. The first three cycles of a PCR reaction. The original DNA is shown in dark gray, the primers in black, newly synthesized DNA in light gray. Each step first involves heating to separate the double strands of DNA, then fast-cooling to allow primers to bind, and finally allowing the DNA polymerase to synthesize a new strand of DNA off of the primer. Dotted lines after the product of cycle 1 shows how the product of one cycle provides the template for the next cycle. The first cycle results in two double-stranded DNA, each composed of an original and the new, second, shorter piece with the primer at one end. The second cycle results in four double-stranded DNA; two are like those after the first cycle. The other two strands are both newly synthesized DNA extended from primers. Because two of these new strands are made off of a strand ending with the other primer, their lengths are determined by where the primers bind. The third cycle, shown only as the final products, results in eight double-stranded DNA molecules; two

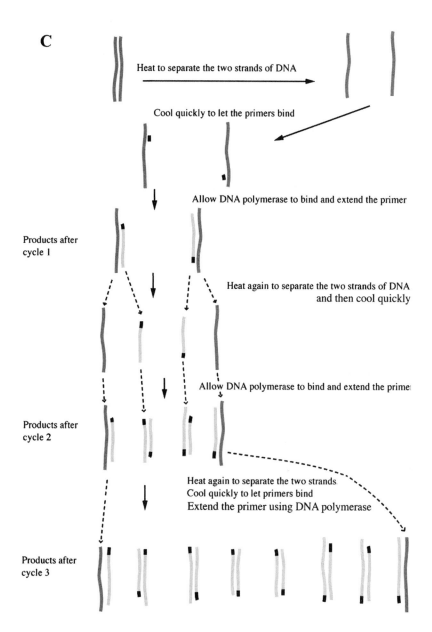

C

Heat to separate the two strands of DNA

Cool quickly to let the primers bind

Allow DNA polymerase to bind and extend the primer

Products after cycle 1

Heat again to separate the two strands of DNA and then cool quickly

Allow DNA polymerase to bind and extend the primer

Products after cycle 2

Heat again to separate the two strands.
Cool quickly to let primers bind
Extend the primer using DNA polymerase

Products after cycle 3

are similar to the results of the first cycle; four are like those just described for cycle 2; and finally, the lengths of two full pairs of strands are defined by the two primers. With each successive cycle after this, the amount of double-stranded DNA doubles. Thus this reaction is called polymerase *chain* reaction. As the cycle number increases, the amount of DNA defined by the two primers increases. This chain reaction allows one to make a great deal of DNA of a specific size and visualize it on a gel.

under those circumstances; once a DNA double helix is completed, the system stops because the DNA is now double-stranded and lacks regions that are partially double-stranded, the necessary condition for DNA synthesis.

This is where Mullis's creativity came to the rescue. He figured that if more primers were available, and if one could separate the two newly formed DNA strands by heating after the first round of replication and then cool the solution rapidly, the newly synthesized DNA would make new bonds with more primers. The DNA polymerase would then make more DNA by using the partially double-stranded regions formed by the primers. Thus by repeating the cycles of heating and cooling, one could make a lot of DNA identical to the DNA that began the process. There was one big problem, however. The DNA polymerase enzyme used to replicate DNA, like most enzymes, was completely destroyed by the heat necessary to separate the double-stranded DNA molecules. Fortunately, biologists had discovered, practically at the same time, that DNA polymerase extracted from the heat-loving bacteria found in hot places *did* resist heating very well. These bacteria live in extreme environments, such as boiling volcanic fumaroles in Yellowstone Park and midoceanic hydrothermal vents.

This is how PCR works: DNA to be copied is mixed with two short, complementary, single-stranded primers, along with nucleotides and heat-resistant DNA polymerase. The solution is heated to separate the DNA strands and cooled rapidly to allow the primers to bind. DNA polymerase then copies the primed DNA strands once. After a few minutes, the mixture is again heated so that the DNA separates into strands, and these are again fast-cooled to allow primers to bind. A second round of replication has taken place. After a few hours, the original piece of DNA has been copied thousands of times. This process is called the polymerase *chain* reaction because at each cycle, the number of DNA molecules is doubled.

The size of the piece that is copied is determined by the region of the original DNA to which the primers are complementary. Since the majority of the product of PCR matches the size of the DNA bracketed by the primers, this piece can be visualized using gel electrophoresis. Gel electrophoresis is a common way to study DNA. It is called "gel" because a thin sheet of Jell-O-like medium is used to separate DNA by size, with the smaller size DNA moving faster while the larger pieces move more slowly. It is called electrophoresis because the DNA moves through the medium due to an electric current run between the ends of the gel. We can then see the DNA pieces by using a stain.

The PCR technique can be used to amplify trace amounts of DNA from drops of dried blood, saliva, or a single hair follicle. It is also used to make DNA from scarce or degraded material, for example, mummies or fossils.

Box 1.1 *PCR and Identification*

We will never forget the horrible events of September 11, 2001, that took place in New York City and Washington, D.C., in which close to three thousand innocent victims were blindly massacred. Yet, thanks to the science of genetics, many surviving relatives have had the solace of knowing that their loved ones' remains were identified. This may bring to many some sense of closure. You probably heard on TV that friends and families of the victims were requested to provide hair- and toothbrushes known to belong to those who died. This is because DNA recovered from a single hair follicle or the very few cells present on a toothbrush can be used to type a person, to obtain this person's "DNA fingerprint."

There is not enough DNA in a hair follicle or a few cells to allow direct genetic typing. However, we have seen that the PCR reaction can amplify DNA samples tens of millions of times. This is what was done in the case of the 9/11 victims where, in many instances, only body parts could be recovered, making other types of identification impossible. DNA was first isolated from hair follicles, for example, and subsequently amplified by PCR with primers known to correspond to extremely variable regions of the human genome. Human DNA contains extensive stretches that do not carry genes. The lengths of these regions vary greatly from individual to individual. Yet these stretches are flanked by other sequences that do not vary much. The principle here is to use primers that bind to the conserved regions and amplify the regions of variable length. This can be done with several sets of primers that amplify a number of different variable regions. The amplified DNA is then characterized by gel electrophoresis.

continued on next page

Box 1.1 *continued*

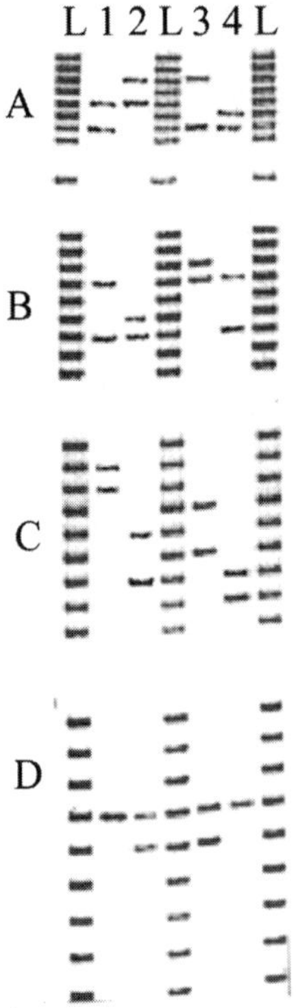

L 1 2 L 3 4 L

A

B

C

D

Figure B.1.1 DNA Fingerprinting. Gel electrophoresis of DNA samples from four individuals, represented by lanes 1–4, amplified using sets of PCR primers to four variable regions in the DNA, labeled here as A, B, C, and D. The lanes labeled L provide identical reference markers used to calibrate the distances that the DNA bands traveled in samples 1–4. Note that only one single band in D is shared by all four individuals.

Next, DNA was extracted from the remains of the victims and processed in the same way. By comparing the length of amplified DNA from the remains with those from the hairbrush, it is possible to provide positive identification of the victims (see figure B.1.1). This application shows that genetic technology is now indispensable to solving forensic problems. The same approach is used in paternity cases.

Another example of DNA typing is of historical importance and has helped uncover an impostor posing as a member of an imperial family. In 1918, Nicholas II Romanov, the last tsar of Russia, his wife Alexandra, and their children were assassinated by Bolshevik revolutionaries. Their bodies were dumped into a shallow unmarked grave. In 1993, their bones were dug up and identified by DNA typing. This was possible because Prince Philip of Edinburgh, the husband of Queen Elizabeth II , is a relative of the deceased Alexandra. Their DNA profiles matched.

Interestingly, shortly after the assassination took place, rumors started circulating that one of the daughters of Nicholas II and Alexandra, Anastasia, had survived. In 1922, a woman claiming to be Anastasia surfaced in Berlin and was able to convince many émigrés of the Russian nobility that she was indeed Anastasia. Some, however, were not convinced. Later, this woman emigrated to the United States under the name of Anna Anderson. She died in 1984.

Box 1.1 *continued*

Thanks to preserved tissue samples that were kept in a hospital where she had undergone surgery, her DNA could be typed. The results were negative; Empress Alexandra and Prince Philip were not related to her. Why Anna Anderson claimed to be Anastasia is unclear. She maintained till her death that she was a Romanov. However, she never benefited, monetarily or otherwise, from her lies. We know now that she was simply an impostor.

Summary

We have learned in this chapter about the chemical composition of DNA, its double-helical structure, and its role as genetic blueprint. We also can see from its double-helical structure, based on complementary base pairing, how DNA can be copied. This understanding of DNA replication has led to the discovery of a technique, PCR, that allows the production of substantial amounts of DNA from very small amounts. PCR is now used on a routine basis in laboratories doing basic research and in forensic laboratories.

Try This at Home: Extract DNA from Vegetables in Your Kitchen

Have you wondered what DNA looks like? It is fairly easy to extract DNA using common equipment and materials found in your kitchen. The following is one recipe for isolating DNA. The recipe mentions onion but you can use other vegetables, such as lettuce or celery.

Ingredients
1 small onion
meat tenderizer
dishwashing detergent
cheesecloth
denatured alcohol (can be purchased at a pharmacy)

continued on next page

Directions

Peel and chop up 1 onion and place in blender.

Add twice the amount of water and blend until fine.

Add 1–2 tablespoons dishwashing detergent. This is to emulsify the membranes around cells that are made of lipids.

Add 1 tablespoon meat tenderizer. Meat tenderizer is typically made from papaya or other fruits that contain protease, an enzyme that breaks down proteins. This helps to release the DNA from proteins.

Gently (so that the mixture does not become foamy) mix in the detergent and meat tenderizer.

Filter through cheesecloth to get rid of plant debris.

Gently layer cold denatured alcohol on top of the clear filtered juice. The white material at the interface is DNA!

If you want to collect the DNA, try spooling it up with a chopstick.

CHAPTER 2

Inheritance of Single-Gene Traits

HUMANS HAVE TINKERED WITH PLANTS and animals since before the dawn of recorded history. For example, it is thought that dogs were domesticated from wild wolves about 15,000 years ago. All the crops we eat today, including wheat, rice, and corn, are the products of thousands of years of breeding for larger grains, edibility, flavor, and other desirable characteristics. In addition to the domestic dog, horses, cows, goats, and pigs were bred from their wild ancestors. Obviously, domestication was a great success, for we now have many varieties of these useful crops and animals suited for different parts of the world and used for different purposes. Yet, despite this obvious success, until about a hundred years ago we did not know how inheritance worked. Breeders could generally state that many traits appeared to be inherited. But when it came to details of that inheritance, some traits seemed to be blends of features from both parents, whereas other traits appeared to copy the trait from only one parent. Sometimes traits from grandparents not expressed in the parents reappeared. Occasionally some useful traits seemed to arise completely anew! How were breeders to make any sense of these observations?

Breeders had trouble understanding how inheritance worked because many traits represent the manifestation of many genes and can also be affected by the environment. Many traits of great interest fall into this category, such as the amount of milk produced by cows, the size of a fruit, the amount of oil in canola seeds, the color of our skin, and our propensity for high blood pressure, just to name a few. We will look into these more complex traits determined by many different genes in chapter 12.

Plants Are Good Organisms for the Study of Inheritance

Near the end of the nineteenth century, many plant breeders were keenly interested in how inheritance worked. They began to realize that what was needed was a systematic study of hybridization, or crossbreeding, between closely related plants possessing obvious differences in a single trait or just a few traits. Fortunately, a serendipitous choice of traits that were under the control of single genes enabled them to make sense of inheritance. Gregor Mendel was the first to figure out how inheritance worked for single-gene traits. His careful choice of plant, peas, and their character, as well as his statistical analysis of many offspring, revealed the bases of inheritance. But his discovery went unappreciated until 1900, long after his death in 1884.

Because many of the early studies of inheritance used plants, let's take a few moments to learn a bit about plant biology and reproduction. Most plants, unlike animals, are hermaphrodites; that is, one individual possesses both male and female parts. The pollen grains are like the sperm cells of the flower. They are contained in a structure called the anther. The ovule is the female reproductive cell that is found at the base of the structure called the style. This structure is topped by a sticky part called stigma that receives the pollen. The fact that plants are hermaphrodites allows scientists to perform controlled crosses that would not be possible with animals. For example, plants can be selfed, short for self-fertilized or self-pollinated. Selfed means that the male (pollen) and female (ovule) contributions to an offspring come from the same individual. Many plants are easily selfed by enclosing the flower to prevent pollen from another individual from entering the flower that contains both the anther and stigma. In order to perform crosses between different individuals, the anther is removed from the individual that will be contributing the ovule. Pollen from the anther of a different individual is picked up with a fine brush and brushed on the stigma of the first individual. This requires a bit of practice, but, once learned, it is fairly easy to do. Also, many plants produce a lot of seeds, that is, offspring, from a single cross. The ease of controlled crosses, the production of many offspring, and relatively fast generation times make many plants ideal for the study of inheritance.

Genes Do Not Blend

Early in the study of heredity, most scientists and breeders thought that traits of offspring were due to blending traits of the parents. After all, there were clear examples of such blending, for example, skin color, the heights of many plants, and so on. This interpretation sounded quite reasonable because male and female gametes contribute to the formation of the offspring. Nevertheless, the idea of blending inheritance could not explain many features of inheritance. One common occurrence that could not be explained by blending inheritance is a trait's skipping generations. That is, a trait is present in the grandparent, not present in either parent, but appears in the offspring.

Crosses of parents that did *not* exhibit this blending feature were seized upon to try and understand inheritance of traits. That is, exceptions to the generally accepted idea of blending inheritance allowed researchers to ask why some traits did not exhibit this behavior. Edith Saunders, at Cambridge University, England, who later became the president of the Genetical Society, conducted extensive studies in the 1890s. She studied a number of different garden and alpine flowers. For example, she noted that there were two varieties of plants in the cabbage family of alpine plants. The plants are similar in all respects, except that one has hairy leaves while the other has smooth leaves. These plants grew right next to each other in the meadow. Saunders reasoned that if insects freely pollinated the flowers, and the traits shown by the offspring were a blend of parental traits, two distinct forms should not persist for long. So, Saunders brought these plants into her garden in order to experiment with them, under conditions where she could control their pollination. Saunders observed that by cross-pollinating a hairy-leafed plant with a smooth-leafed one, she obtained twenty-one hairy plants and ninety-nine smooth-leafed plants as the offspring (figure 2.1) and no plants with intermediate hairiness. In another example, she used a fragrant garden flower, called stock, that has very smooth leaves. Crossing this smooth-leafed plant with one covered in gray hair from the wild resulted in only smooth or completely hair-covered offspring and nothing in between.

smooth   hairy

99 smooth
21 hairy

Figure 2.1 Cross Between Smooth-Leafed Plant and Hairy-Leafed Plant. When Edith Saunders crossed smooth-leafed plants with hairy-leafed plants, she obtained only smooth-leafed or hairy-leafed plants. No partially or less hairy leaves were observed. These are images photographed through the microscope illustrate the difference between hairy and smooth leaves. In a cross between these plants, Saunders obtained ninety-nine smooth-leafed plants and twenty-one hairy-leafed plants. Photos of leaves courtesy of Valerie Lynch-Holm. The plants are from the WSU Owenby Herbarium.

Saunders interpreted her results as clear examples of nonblending inheritance. The stage was set for the rediscovery of Mendel's results.

Rules of Inheritance

The Dutch botanist Hugo De Vries rediscovered Mendel's work in 1900. He looked at many different flowering plants, but for each cross, he focused on only a single trait that differed between closely related plants. When De Vries crossed a plant of one type with another plant differing by only a single trait, all the resulting offspring showed the trait of one parent. For example, he crossed a red campion with a white campion. He found that the first generation seedlings all produced red flowers (figure 2.2). When he then allowed these red flowered offspring to self-pollinate, he got 73 percent red and 27 percent white, or close to a 3 to 1 ratio. He did another experiment using the same species of plants, but this time he chose varieties that differed in the hairiness of their leaves. The first generation produced all hairy plants, but a self-pollination of those hairy-leafed offspring produced 72 percent hairy-leafed and 28 percent smooth-leafed plants. Among many such experiments, he saw approximately 75 percent of one type

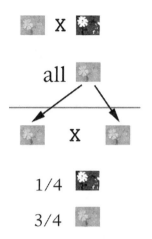

Figure 2.2 Cross Between Red and White Campion Flowers. The cross-pollination of a red campion with a white campion is depicted at the top. De Vries obtained all red campions from this cross. Then, he self-fertilized, or self-pollinated, these red campions with their own pollen. From this cross, shown below, he observed 25 percent white and 75 percent red campions.

and 25 percent of the other type: a 3 to 1 ratio. From these and other results of his experiments, using a number of different plants, he came up with the following explanations:

- Each parent provides equal genetic contribution to the off-spring, so the offspring has two copies of their genetic contribution, one from the mother and the other from the father.
- When considering a single characteristic, one expression is dominant (visibly present) and the other is recessive (not visible). This is the rule of "Dominance."
- In a plant, these characteristics are combined, but when that plant makes pollen grains or ovules, these characteristics are separated, or segregated, so that each grain of pollen and each ovule retain only one of the two characteristics. This is the rule of "Segregation."

Behavior of Chromosomes

By the time of the rediscovery of Mendel's work, biologists could stain DNA and had observed that DNA was found in chromosomes. The discovery of chromosomes led to the recognition of two types of cell division. One type, called mitosis, resulted in two daughter cells that had the same number of chromosomes as the parent cell (figure 2.3). Another type of cell division was observed in the formation of gametes, eggs or ovules and sperm or pollen. In the process of making gametes, through a type of cell division, called meiosis, each pair of chromosomes is separated into different gametes. The four resultant cells each have half the number of chromosomes of the parent cell (figure 2.3). These observations of the behavior of chromosomes provide a way to explain the results of the cross of plants differing in a single character. Figure 2.3 shows the simple example of fruit fly with only four pairs of chromosomes. In contrast, humans have twenty-three pairs and peas, for example, have seven pairs of chromosomes. The fruit fly has been and is still used extensively in genetic studies. We will see examples of this use in this chapter and later in the book.

The Punnett Square
Observing the behavior of chromosomes provides a way to explain the results of crossing plants that differ in a single character. That is, in meiosis (see figure 2.3) the two copies of the chromosomes are segregated into different gametes. When a new individual is formed by the fertilization of an ovule with a pollen grain, both pollen and ovule contribute a set of chromosomes to form the new individual. This interpretation of the behavior of chromosomes during meiosis, combined with the inheritance of characteristics, can be conveniently illustrated by the use of the Punnett square, named after the biologist Reginald C. Punnett (figure 2.4). Figure 2.4.A is an example of a Punnett square showing the chromosomes of an individual. Each parent provides one set of chromosomes in each gamete, egg or sperm. These are shown at the top and left of the square. Then, in the process of making a new individual, the gametes from the parents join to form chromosome pairs in this new individual. The four boxes indicate that four types of offspring appear in equal proportions, that is, one each. Early plant breeders thus understood important features of inheritance and their physical basis without knowing about genes and DNA. Figure 2.4.B shows a Punnett square with symbols for hairy-leafed versus smooth-leafed plants, as in the experiment done by De Vries. The appearance of traits in an organism, whether hairy or smooth or red or white, is called the "phenotype." This is in contrast to the genetic composition of the individual, called the "genotype." By convention, capital letters indicate "dominant" characteristics and lowercase letters indicate "re-

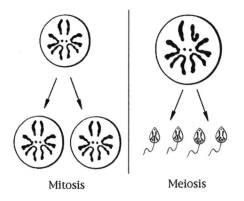

Mitosis   Meiosis

Figure 2.3 Types of Cell Division. Cells have two possible different types of division. In these examples, both begin with one cell equipped with four pairs of chromosomes. Note in this example that there are two tiny, dotlike chromosomes and three additional pairs of chromosomes. One type of division, depicted on the left, is called mitosis, in which one cell doubles the number of chromosomes and then divides them equally into two daughter cells, producing two identical cells. The second type of division is called meiosis. It is shown on the right. Meiosis produces eggs and sperm. This process also begins by doubling the number of chromosomes but then divides them into four cells, each with half the number of chromosomes of the original cell. These examples use the chromosome configuration of the female fruit fly on the left and male fruit fly on the right. Note the hook shaped Y chromosome in the male cell. The chromosomes in both are actually the same size and much smaller in relation to the cell than shown.

cessive" characteristics. The Punnett square in Figure 2.4.B shows De Vries's first cross between a hairy-leafed parent, shown at the top, and a smooth-leafed parent, shown to the left. All the offspring have one each of **H** and **h** genes. Since we know that all the offspring have hairy leaves, the combination of **H** and **h** must result in hairy leaves. Thus **H**, the hairy-leaf trait, dominates over **h**, the smooth-leaf trait. Then, in Figure 2.4.C, these hairy-leafed offspring are allowed to self-pollinate. The Punnett square shows that the parents' genotypes are the same since the pollen and the ovules come from the same plant and have one **H** and one **h** each. In this case we get one-quarter **HH** offspring, which, we know from the original hairy-leafed parent, are hairy, and we also get two-quarters or one-half **Hh**, which are like the hairy offspring of the first cross and the hairy parent of this cross. Finally we get one-quarter **hh** offspring that have smooth leaves. Thus, what is called the genotype, represented by the bold letters, the composition of genes, determines the appearance of the leaves, or the phenotype. If there is at least one dominant gene, that gene determines

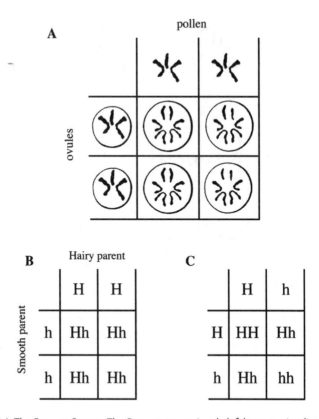

Figure 2.4 The Punnett Square. The Punnett square is a helpful way to visualize what we would expect from a cross. A. A Punnett square is illustrated using fictitious chromosomal compositions of pollen grains (top) and ovules (left). Each gamete, ovule, or pollen has three chromosomes. The four types of offspring resulting from the combination of pollen with ovules are shown in the four boxes. B. DeVries's first cross between a hairy-leafed plant and a smooth-leafed plant. Note that all offspring are heterozygous with genotype **Hh**. C. The heterozygous **Hh** offspring from the cross in B are used as parents in this cross. Note that we expect to get 75 percent hairy and 25 percent smooth-leafed plants.

the phenotype. If both copies of the gene are recessive, then the trait that is determined by the recessive gene is observed in the phenotype. In other words, a recessive phenotype can *only* be observed if *both* copies of the gene are recessive. This type of analysis can be done for the red and white flowers, and one comes out again with 75 percent red and 25 percent white, or a 3 to 1 ratio.

In cases of single-gene traits where two different forms of the gene exist, we end up with genotypes represented by combinations of upper- and lowercase letters. Because we encounter this situation often, let us now introduce new terms to describe genotypes more precisely. For this, we will use a final example from De Vries's work, poppies with either a black spot or white spot at the base of the petal. As before, a cross between the two produced all black-spotted plants, showing that the black-spotted characteristic was dominant over the white-spotted one. When De Vries then allowed those black-spotted poppies to self-pollinate, he obtained 75 percent black-spotted and 25 percent white-spotted, as expected. He then saw that among the black-spotted ones, there were two types that could be distinguished by self-pollination. Of the 25 percent, he obtained all black-spotted plants upon self-pollination. With the remainder he again obtained 75 percent black and 25 percent white. These results confirmed the previous ones obtained with the hairy and smooth leaf characteristics. Individuals that carry two identical genes (for example, **HH** or **hh**) are called "homozygotes" (adjective = homozygous). Individuals that carry two different genes (for example, **Hh**) are called "heterozygotes" (adjective = heterozygous).

Carl Correns, another rediscoverer of Mendel's work, wrote in 1900:

> I thought I had found something new. But then I convinced myself that Abbot Gregor Mendel in Brunn, had, during the 1860's, not only obtained the same results through extensive experiment with peas . . . as did De Vries and I, but also given the same explanation. (C. Correns, "G. Mendel's Law Concerning the Behavior of Progeny of Varietal Hybrids," reprint, *Genetics* 35, no. 2 (1950): 33–41.)

Correns made another important point, that not all characteristics follow the rules of dominance and segregation. Indeed, even among those plants that he and others used and that follow these rules of dominance and segregation, there were other characteristics that appeared to be a blend of the parents. So why do we discuss rules of inheritance when they do not seem to be universal? We will see that the rules of inheritance, now called Mendelian inheritance, are the simplest manifestations of heredity because they correspond to the actions of single genes. Once we fully understand this simplest form of inheritance involving single genes, we will be able to tackle more complicated forms that involve many genes (see chapter 12).

Incomplete Dominance

One complex type of inheritance that may appear as blending inheritance is called "incomplete dominance." An example of this is seen in the four-o'clock plant, named so because its flowers open toward evening. This plant, like the campion, can be found in red and white varieties. However, unlike the campion, when a red flower is crossed with a white flower, we get pink flowers! This seems like a perfect example of blending. But what happens when we now self-pollinate the pinks? We might expect to get all pink flowers, since the flower colors of the parents seem to blend to form the flower colors of the progeny. Instead, we get 25 percent white, 50 percent pink, and 25 percent red. Thus, clearly this is *not* an example of blending inheritance. Let us analyze this result in the same way as we did the campion flowers. In that case, red-flowered offspring of crossing red flowers with white flowers could be represented by the **Rr** genotype. Self-pollination of these heterozygous red flowers **(Rr)** produced 25 percent white and 75 percent red. But we also realize that 50 percent of these red flowers were heterozygous **(Rr)** and 25 percent were homozygous **(RR)**. We can understand the basis for inheritance in four-o'clock flower color if the heterozygous individuals **(Rr)** are pink rather than red (figure 2.5). These are pink because the dominant **(R)** form of the gene is weakly expressed and it takes two copies of **R** to get red.

Mendel and the rediscoverers of his laws of genetics were particularly fortunate in their choice of organism: plants. Indeed, in the overwhelming majority of plants, the same individual carries the pollen grains, the male gametes, and the ovules, the female gametes. Therefore, when gametes are formed, both pollen grains and ovules will have a similar genetic composition. For example, if the plant is homozygous **AA** or **aa**, all gametes will contain a copy of **A** or **a**. *If the plant is an Aa heterozygote, half the pollen grains will carry an A and the other half will carry an a. This also holds true for the ovules.* Such a situation very rarely occurs among animals because the two sexes are usually differentiated and the genetic composition of their gametes is harder to determine. Nevertheless, as we will see in the next section, later geneticists have taken advantage of sex differentiation in animals to further our understanding of genes and chromosomes.

	R	R
r	R r	R r
r	R r	R r

	R	r
R	R R	R r
r	R r	r r

For campion flowers, Rr is red
For four-o'clock flowers, Rr is pink

For campion, 3/4 red and 1/4 white
For four-o'clocks, 1/4 red, 1/2 pink, and 1/4 white

Figure 2.5 Incomplete Dominance Illustrated by the Four-o'clock Flower. We represent red flowers with **RR** and white flowers with **rr**. Offspring are all **Rr**. In the case of the four-o'clock flower, **Rr** is not red but pink. We can see that this is *not* a case of blending inheritance because when we self-pollinate the pink flowers, we get back one-quarter **RR**, which are red, one-half **Rr**, which are pink, and one-quarter **rr**, which are white. Thus we observe 75 percent *colored* flowers and 25 percent *white* flowers, as in de Vries's experiment with campions.

Sex Is Also Determined by Inheritance Rules

In this section we will see how the chromosome theory of inheritance was put on firm footing thanks to the study of genes located on those chromosomes that determine the sex of an individual. We will also see that, in many animals, specific chromosomes called sex chromosomes determine sex.

In 1910, Thomas Hunt Morgan of Columbia University in New York demonstrated that genes reside on chromosomes. It was known then that males and females of the fruit fly *Drosophila melanogaster* had different chromosomes. Fruit flies, though small, can be easily sexed by observing them under the microscope. The males have a rounded end to their abdomen while the females have a pointed end. The sexes both have three pairs of similar chromosomes, but females contained an extra pair of chromosomes called X chromosomes while males contained a single X chromosome paired with a chromosome that looked different from the X, called the Y chromosome (see figure 2.3). Thus, the Y chromosome is not found in the female fruit fly.

Fruit flies usually have red eyes, but one day Morgan spotted a white-eyed male fly in the midst of his large fly collection. Intrigued by this difference, he mated this male with normal, red-eyed females. The offspring generated all red-eyed flies. Morgan's conclusion was that red pigmentation was dominant over white in eye color. He then did what Mendel, De Vries, and others had done before him: he allowed these first-generation flies to mate and produce a second generation (this is equivalent to self-pollination in plants). He expected to get a 3 to 1 ratio of red-eyed flies to white-eyed flies. To his surprise, Morgan observed that *no* white-eyed females appeared in this second generation, although red-eyed males and females as well as white-eyed males did appear. Thus there was a difference in the inheritance of the eye-color trait among males and females.

Taking this difference in the results among males and females as a cue, Morgan hypothesized that the trait was located on the sex chromosomes, which, he knew, differed between males and females. If, for example, the trait was only found on the X chromosome, we can try to explain the above result using the Punnett square. Let us designate **R** as the dominant gene determining red eye color and **r** as the recessive gene determining white eye color. The white-eyed male would be $X^r Y$, with only one copy of the recessive gene, and the red-eyed females were $X^R X^R$ homozygous dominant. During meiosis, the chromosomes segregate and yield gametes that contain one X^r and one **Y** (for the male), or gametes containing a single X^R (for the females). Note that the Y chromosome in this hypothesis carries no **R** or **r**. If we arrange these gametes as in figure 2.4, we get the Punnett square shown in Figure 2.6.A. That is, if the gene for eye color was on the X chromosome and the white-eye trait was recessive, no offspring would show a white-eye phenotype. Because red color is dominant, heterozygous females would be red eyed and males would be red eyed as well because they carry the dominant **R** gene only.

Then Morgan crossed these offspring with each other, that is, the red-eyed male with the heterozygous red-eyed female. Among the offspring, he obtained both red-eyed and white-eyed males, but only red-eyed females. This result can be explained using the Punnett square shown in Figure 2.6.B. That is, the red-eyed males ($X^R Y$) and heterozygous red-eyed females ($X^R X^r$) would produce both red-eyed

(X^RY) and white-eyed (X^rY) males, but only red-eyed (X^RX^R and X^RX^r) females. To test this hypothesis of genes' residing on sex chromosomes, Morgan mated his original white-eyed (X^rY) male with the first-generation daughters, which all had red eyes (X^RX^r), as we saw. He used the genotypes derived from his analysis and predicted that this cross should produce red-eyed and white-eyed offspring of both sexes as shown in Figure 2.6.C. Indeed, there were equal proportions of red-eyed females, white-eyed females, white-eyed males, and red-eyed males among the offspring of this cross. Thus, the recessive white-eyed trait could, after all, exist in females. What's more, the four categories of flies were present in roughly equal numbers in the offspring. From this, Morgan derived the following: *Traits determined by genes on the X chromosome are always transmitted from mothers to sons, since, to be sons, they must get their Y chromosomes from their fathers. The single X chromosomes in males must come from their mothers. The Y chromosome does not carry the genes that are on the X chromosome, so whether the gene on the X chromosome is dominant or recessive, it cannot be masked by another copy of the gene on the Y chromosome. In males, any gene present on the X chromosome is always expressed.* Indeed, in all the cases where males have two different sex chromosomes, called in this case the "heterogametic" sex (XY), and the females have two of the same sex chromosomes, that is, the "homogametic" sex (XX), fathers do not contribute an X chromosome to their male offspring; they contribute only a Y chromosome. Mothers, however, contribute X chromosomes to both sons and daughters, while fathers only transmit an X chromosome to their daughters. This is exactly what happens in humans as well.

Morgan thought that if his hypothesis was correct he could perform another cross in which he knew the genotype of the parents from the phenotype and he should be able to predict the results of the cross. This was the case with a cross involving a white-eyed female and a red-eyed male. Each of these two parents could only have one genotype, that is X^rX^r and X^RY. Morgan predicted that he would get all red-eyed females and all white-eyed males. The result of this cross gave offspring that fully confirmed his original hypothesis. *You should build a Punnett square for this cross and convince yourself that Morgan was right.* Thus, in one fell swoop, Morgan demonstrated that a trait (eye color) could be associated with a chromosome (the X chromosome). In addition, we understand why sex ratios in fruit flies, and in

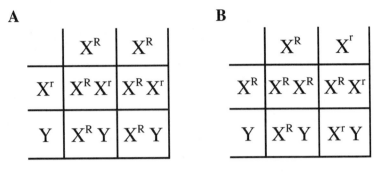

Figure 2.6 A Sex-Linked Trait Illustrated by Red (Normal) and White (Mutant) Eye Color of Fruit Flies. A. When Morgan crossed a white-eyed male with a normal red-eyed female, he got all red-eyed flies. Note that each of the four offspring has one copy of X^R. B. In the next cross, Morgan crossed the red-eyed female and red-eyed male offspring from the previous cross with each other. Their progeny are red- and white-eyed males, but the females all have at least one **R** so are all red-eyed. C. Finally, he crossed the original white-eyed male with the red-eyed females that he produced in the first cross. This resulted in equal number of red-eyed and white-eyed females and males.

other animals whose sex is determined by sex chromosomes, are 50 percent males and 50 percent females; the Punnett square readily demonstrates why (see also figure 2.6). Therefore, Mendelian traits do indeed reside on chromosomes, and specific chromosomes determine sex, also in a Mendelian fashion. Innumerable crosses in a large number of species have confirmed this.

Now, not all life forms rely on sex chromosomes to determine sex. In plants, there are no sex chromosomes that differentiate male from female. In some insects and worms, females are XX and males have only one X chromosome and no Y chromosome. In birds and fishes, males are the homogametic sex, that is, they have two identical sex chromosomes, and females are the heterogametic sex, with two different sex chromosomes. However, in some reptiles, it is the incubation temperature of eggs that determines the sex of the developing embryo. Some species produce mostly females at high temperature, while at low temperature males predominate. In other species, more males are produced at intermediate temperatures and females are produced at high and low temperatures. This is due to the effect of temperature on hormone production in these animals. Thus there are a variety of ways in which Mother Nature determines sex in different organisms.

Summary

This introduction to the rules of inheritance has relied on simple cases where a trait is determined by a single gene. In this particular context, one form of the gene, the dominant form, confers an easily visible trait (phenotype) to an organism. The other form of the gene, the recessive form, only confers a different trait when it is present in a homozygote. Thus, the dominant form "masks" the recessive form in a heterozygote. We have also seen that genes reside on chromosomes and that individual organisms contain pairs of chromosomes. The mother contributes one half of the pair, and the father contributes the other half. In some organisms, such as fruit flies and humans, sex is determined by chromosomes. The segregation of sex chromosomes explains the 1 to 1 sex ratio of these organisms.

CHAPTER 3

Mendelian Traits in Humans

WE NOW KNOW THAT THE BEHAVIOR of simple traits, determined by single genes that have a dominant and a recessive form, can be understood using the Punnett square. Yet we saw that Carl Correns, one of the rediscoverers of Mendel's work, realized that even in a same species of flowers that seemed to follow Mendel's rules for some traits, not all traits exhibited this simple form of inheritance. Many geneticists wondered whether these simple rules applied at all to animals and to humans. We will see in this chapter that simple rules of Mendelian inheritance do indeed apply to humans.

It turns out that by looking hard, biologists have come up with a few easily visible human traits that are determined by single genes. Some quite odd traits are reported to behave according to Mendel's rules. These include the widow's peak hairline, the ability to curl one's tongue, attached earlobes, and extra fingers. However, other Mendelian traits in humans are of medical importance, such as blood type, color blindness, and certain metabolic diseases. Let us study each of these traits in a little more detail.

Blood Types

First, let us examine the case of blood types. The A, B, and O blood types provide a clear example of a critically important Mendelian trait because being transfused with the wrong blood can be lethal. Blood types are subdivided into the A, B, AB, and O groups. Single inherited genes determine these blood types. Thus, there is a form of the gene for A blood type, and a different form of the gene confers the B blood type. These genes produce material on the outside of the red blood cells. There is also a null form, called O type, that does not make either A or B products (figure 3.1). The O blood type is known as universal donor, because O red blood cells do not produce any A or B compounds and thus cannot cause clotting. The AB blood type is known as universal recipient because both A and B compounds are present. Type A people can only be transfused with A or O blood while type B people can only receive B or O blood. Remember, we get one gene from our mother and one gene from our father. For example, we can get an **A** form from both our mother and father, in which case we will have two copies of **A** and have an A blood type. The same happens with two copies of **B**, giving a B blood type, whereas the null form that results in O blood type is obtained when both parents contribute the **O** form of the gene. If we get an **A** gene from one parent and a null from the other parent, we still will be A blood type but we would have an **AO** genotype. The same happens with B, a **BO** genotype giving a B phenotype. In this case, **A** or **B** is dominant over **O**. If we receive an **A** gene from one parent and a **B** gene from the other parent, we will have an AB blood type. This last case shows that **A** and **B** are neither dominant nor recessive to each other; both are expressed. This is an example of "codominance," since both **A** type and **B** type contribute equally to the phenotype when both forms are present.

Sex-Linked Traits: Hemophilia

We have already seen that in humans and many other animals, sex is determined by sex chromosomes. As we saw in chapter 2, one such trait is eye color in fruit flies. In humans, as in fruit flies, females have XX sex chromosomes, and males have XY sex chromosomes. Scientists have discovered that the X chromosome carries many more genes than the Y chromosome and the Y chromosome is very small.

Hemophilia in humans is a classic example of a Mendelian trait that resides on the X chromosome. Individuals with hemophilia suffer excessive bleeding caused by the inability of their blood to clot. Like many diseases, however, hemophilia exists in different forms. There are actually three different diseases categorized as hemophilia.

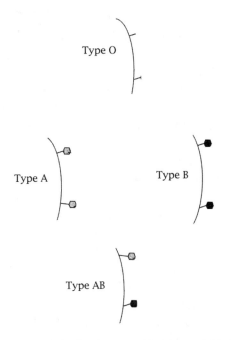

Figure 3.1 Differences Between Blood Types. A simplified diagram of the difference between A, B, and O blood types showing the differences in the surface sugars of red blood cells. The O blood type (top) has a set of sugar molecules shown as dashes. The other blood types have an additional sugar attached. A person with two copies of the **A** or **B** blood type gene is indistinguishable from someone with an A and an O sugar, and vice versa with the B blood type. An AB individual has both an A-type sugar and a B-type sugar. This diagram greatly exaggerates the size of the sugar molecules with respect to cell size. In reality one would not be able to see the sugars even under a microscope.

Two of those forms, called hemophilia A and hemophilia B, are due to defects in genes located on the X chromosome.

Let us see how the pattern of inheritance is affected when a gene is located on the X chromosome. The classic example from history is that of Queen Victoria of Great Britain and her descendants. Queen Victoria was a carrier of hemophilia. This means that she had one normal copy of the blood clotting gene and one defective copy. The form of the gene responsible for the hemophilia trait is recessive to the normal gene, so Queen Victoria did not show any symptoms of the disease. Her husband, Prince Albert, did not exhibit the disease either. Males have an X and a Y chromosome and the Y chromosome does not carry the genes for blood clotting at all. Since recessive traits on the X chromosome of males are expressed because there is not another X chromosome to mask a recessive trait, he could not have carried the hemophilia gene.

The Punnett square in figure 3.2.A shows that Queen Victoria's sons had a 50 percent chance of being afflicted with hemophilia, and her daughters had a 50 percent chance of being carriers themselves. Thus, males inherit hemophilia from their carrier mothers. Daughters from a carrier mother and a normal father have a 50 percent chance of being carriers themselves.

Sex-Linked Traits: Color Blindness

Another example of a trait located on the X chromosome is the most common form of color blindness, the type that prevents afflicted individuals from distinguishing between red and green. A color-blind man can only have one copy of the gene for the color-blindness trait, indicated by **r** in the Punnett square in figure 3.2.B, because he has only one X chromosome. He is color-blind because there is no copy of that gene—normal or abnormal—on his Y chromosome. If a color-blind man has children with a woman who does not carry a color-blind trait, one can predict that none of their children will be color blind, as shown in the Punnett square in figure 3.2.B. So where do color-blind men come from? We can use the Punnett square to work in reverse and solve the problem. First, we can fill in the Punnett square with information that is shown in figure 3.2.C. We know that one parent must be male and has a Y chromosome. In addition, half the children are male and carry the Y chromosome. We also know that at least one son is color-blind. Therefore, he must have the color blindness gene **r** on his X chromosome. As we know, the son inherited his X chromosome from his mother. This means that his mother must have at least one copy of the recessive gene. Figure 3.2.D shows an example in which both parents are not color blind, indicating that the father has one normal copy of the gene on his X chromosome but that the mother is a carrier. *Can you use another Punnett square and determine under what circumstances a colorblind girl will be born?*

Prostate and Breast Cancer

There are many other cases of genetic conditions due to a single-gene defect. You can search for these on the Online Mendelian Inheritance in Man (URL in appendix A). A great majority of these diseases are very rare. And a great many of these genetic diseases have nongenetic

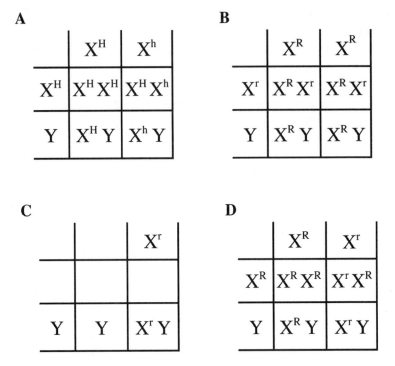

A

	X^H	X^h
X^H	$X^H X^H$	$X^H X^h$
Y	$X^H Y$	$X^h Y$

B

	X^R	X^R
X^r	$X^R X^r$	$X^R X^r$
Y	$X^R Y$	$X^R Y$

C

		X^r
Y	Y	$X^r Y$

D

	X^R	X^r
X^R	$X^R X^R$	$X^r X^R$
Y	$X^R Y$	$X^r Y$

Figure 3.2 Punnett Squares Used to Calculate the Probability of Inheriting Diseases Determined by Genes Located on the X Chromosome. A. This Punnett square shows the probability of transmission of hemophilia, an X-linked trait. The hemophilia gene is only found on the X chromosome and not on the Y. **H** indicates a normal gene, while **h** indicates the hemophilia form. A carrier, that is, a female with one **H** and one **h** on her X chromosomes, who produces children with a nonhemophilic man has 50 percent chance of passing the genes to both her sons and daughters. However, because daughters also get a normal **H** on the X chromosome from their father, 50 percent of them (with the **h** gene) are carriers. Sons only get their one X chromosome from the mother, so 50 percent of them (with the **h** gene) exhibit the hemophilia trait. B. This Punnett square shows the expected ratio of children with a color-blindness trait in a mating where only the father carries the color-blindness trait. This trait is indicated by **r**, while **R** indicates normal. Because the X chromosome of the color-blind father can only be passed to his daughters, and remembering that they get a normal copy of the gene from their mother, the daughters are heterozygous carriers of the color-blindness trait but are not color-blind. All the sons receive their X chromosome from their mother and do not inherit the color-blindness trait, so have normal color vision. C. A partially filled-in Punnett square where we know that at least one son is color-blind. The symbols are as in figure 3.2.B. D. A completely filled-in Punnett square where at least one son is color-blind, but neither of his parents is color-blind. Note that the mother must be a carrier or heterozygous for the trait.

counterparts. Cancer is an example of a trait that is sometimes caused by a single gene defect, and sometimes caused by other factors. Two types of cancers are now known to have a genetic basis: prostate cancer and breast cancer.

Let us first examine the case of prostate cancer. There is a single dominant form of a gene for hereditary prostate cancer called HPC1. This form of the gene greatly increases the chance that an individual carrying a single copy of that gene will get prostate cancer. Although prostate cancer is the most common cancer among men in the USA, the defect in HPC1 probably accounts for only 5–10 percent of all prostate cancers. Other cases of this disease are due to other cancer-susceptibility genes and to environmental influences.

The situation is similar with breast cancer in women. Scientists have discovered defects in two genes that greatly increase the chance of a woman getting breast cancer. These defects occur in genes called BRCA1 and BRCA2 (for breast cancer). If a woman has a particular defective copy of either of these genes, she has a greatly increased probability of getting breast cancer at an early age. The cumulative lifetime risk of getting breast cancer is 92 percent for women with a copy of the defective gene while the cumulative lifetime risk for a woman without this defective gene is 10 percent. Yet because there are many more women who do not have a defective version of these genes, the percentage of all breast cancer patients that have a defective copy of the gene is only about 5–10 percent. As in the case of prostate cancer, other cancer-susceptibility genes can be involved, as well as environmental factors.

Genetic Metabolic Diseases

There is a growing list of diseases attributed to defects in single genes. We will use examples of such diseases to show that Mendelian inheritance indeed applies to people. Once an observant mother noticed a "mousy" smell in one of her children's diapers. This led to the discovery of phenylketonuria in the 1930s. The mother brought her observation to her doctor's attention. He eventually determined that the smell was due to an unusually high amount of a metabolic byproduct of the amino acid phenylalanine. This byproduct is present in the urine of all individuals afflicted with phenylketonuria, or PKU for short.

Individuals with PKU lack the enzyme called phenylalanine hydroxylase, which converts the amino acid phenylalanine to another essential amino acid, tyrosine. This in and of itself is not harmful because people normally get enough tyrosine from their diet. In PKU patients, phenylalanine is not converted into tyrosine; it accumulates and is metabolized into toxic byproducts. These byproducts cause mental retardation. Because one normal copy of this gene encoding phenylalanine hydroxylase is sufficient for function, two defective copies are necessary for the PKU trait to show; thus PKU is a recessive trait. Infants with PKU appear normal at birth since, while in their mother's womb, the mother's enzymes take care of the excess phenylalanine present in the fetus. If detected early, PKU individuals can benefit from dietary therapy, which involves reducing the amount of phenylalanine in their diet. Mental retardation can be avoided if this therapy is strictly followed. Today, all fifty states test newborns for PKU.

Box 3.1 *Warning on Diet Products*

Have you noticed the warning on diet products, such as diet pop cans that read "Warning, phenylketonurics, contains phenylalanine," and wondered what it meant? Some may have wondered if the soda causes cancer. No, the warning is not because it causes cancer. Phenylketonurics refers to individuals with a genetic disease called phenylketonuria, or PKU. These individuals cannot consume normal quantities of the amino acid phenylalanine because they have a genetic condition that prevents their bodies from getting rid of excess phenylalanine. The sweetener found in these diet products is called aspartame and goes under the brand names Nutrasweet or Equal. Aspartame is a chemical made up of two amino acids, aspartic acid and phenylalanine. Aspartame is broken down in our bodies to its amino acid components, thus we get phenylalanine in our diets from consuming aspartame. A 12 ounce can of diet soda contains 180 milligrams of aspartame. Although this quantity is quite small, PKU individuals cannot get rid of excess phenylalanine in their diet. The excess phenylalanine is degraded into a toxic product that can cause irreversible neurological damage, thus PKU individuals should not consume aspartame.

So now you know what the warning means and why it is there. PKU is the first genetic disease for which we have instituted mass

Box 1.1 *continued*

newborn screening. Now all fifty states test for PKU in newborns. The early detection allows affected families to treat PKU babies with special diets low in phenylalanine so that they can lead normal lives. Each year, several hundred PKU individuals are born in the United States, and it is estimated that there are thousands of PKU individuals now living in the country.

Another similar metabolic disease is galactosemia. This again is a recessive trait. It is characterized by the inability to convert the sugar galactose to glucose. Lactose, the sugar present in milk, is composed of two chemically linked sugars, galactose and glucose. When ingested, enzymes break down lactose into galactose and glucose. Thus, babies who drink mother's milk or cow milk metabolize lactose into glucose and galactose. This galactose is subsequently metabolized into glucose. Babies' cells use glucose as a powerful energy source. Thus, lactose is a great nutrient for normal babies. However, for babies with galactosemia, lactose is fatal! This is because they cannot metabolize galactose, which then accumulates and reaches toxic levels. If this condition is unrecognized and left untreated, babies begin to vomit and suffer from diarrhea within a few days, or at most a few weeks, after birth. Eventually, they die. The treatment for galactosemia, as in the case of PKU, is a relatively simple dietary change. If affected babies avoid galactose, they will survive fairly normally.

Sickle-cell Anemia

Sickle-cell anemia is a disease found to be more common among Africans and individuals of African descent. This disease is caused by a defect in the protein hemoglobin, which is responsible for the red color of red blood cells and carries oxygen in our blood. Normal red blood cells rush from the lungs through our arteries all the way to the periphery of our bodies, single file through tiny capillaries, and then back, through the veins, to the heart. These red blood cells are flat, disc shaped, and can somewhat deform, which allows them to squeeze through the capillaries. In individuals with sickle-cell anemia, the defective hemoglobin crystallizes into a sickle shape and deforms the red blood cells, thus giving the disease its name. This abnormal shape of the red blood cells impedes the proper flow of blood, causing painful "sickling" events. Sickle-cell anemia is a recessive trait. Thus the genotype of an affected child is **aa**. The Punnett square in figure 3.3.A–B demonstrates how to determine the genotype of the parents of an affected child. If the parents are normal, we can "backfill" the Punnett square as shown in figure 3.3.B. If you complete this Punnett square you will see that the probability of two carriers, that is, heterozygous, normal, individuals, having an affected child is 25 percent. This is a devastating disease, with only marginally effective therapies. Sickle-cell anemia, along with other diseases affecting the hemoglobin gene, is tested for among newborns in many states.

Hemochromatosis

One of the most underdiagnosed genetic diseases is hemochromatosis, also called iron-overload disease. This trait is recessive. The condition leads to excessive accumulation of iron and basically causes "rusting of the organs." This trait appears to be much more common than previously thought, with approximately 1 in 200 to 400 individuals affected, and approximately 1 in 10 are carriers. The multitude of symptoms of this disease, from cirrhosis of the liver to cancer, heart problems, diabetes, and arthritis, makes accurate diagnosis, based only on phenotype, difficult.

There is now a reliable test for hemochromatosis. What is more, the treatment is simple, though it may seem old-fashioned or even gross. It consists in bloodletting, or purposely removing blood. The treatment works because blood is an iron-rich fluid. So if you are an affected person, donate blood often! Alternatively, some have opted for a treatment involving leeches that, as we all know, are blood-sucking animals. If therapy is begun before organ damage is detected, the affected individual can expect a long life. But this means that the condition must be detected before damage is done. Because it is now recognized to be the most common of the known genetic diseases, healthcare professionals are weighing the merits and cost of screening people for this disease.

An interesting aspect of hemochromatosis is the observation that more males than females are affected. At first glance, we might think that the defective gene is located on the X chromosome and that hemochromatosis is thus a sex-linked trait. Yet this turns out not to be.

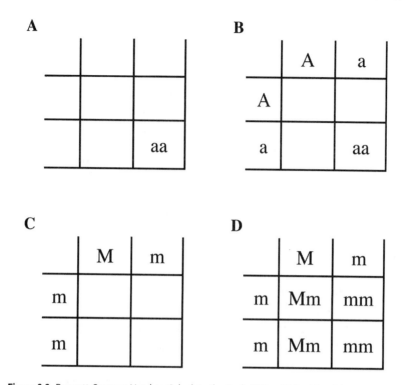

Figure 3.3 Punnett Squares Used to Calculate the Probability of Inheriting Diseases Determined by a Single Gene Not Located on the X Chromosome. A and B show the steps in determining the genotype of normal parents who have a child with sickle-cell anemia, a recessive disease. **A** is the normal form of the gene and **a** its abnormal form. C and D show the steps in determining the probability of having a child with Marfan syndrome, a dominant disease. Here, **M** is the abnormal form of the gene and **m** its normal form. A. A partially filled-in Punnett square showing only the child with sickle-cell anemia (genotype **aa**) in one of the four offspring squares. B. A partially filled-in Punnett square that shows that both parents must be heterozygous (**Aa**) if they are normal and have a child with sickle-cell anemia. C. A partially filled-in Punnett square showing only the heterozygous parent with Marfan syndrome (**Mm**) and the normal parent who must be homozygous normal (**mm**) because Marfan syndrome is a dominant trait. D. A completely filled-in Punnett square that shows a 50 percent chance for this couple to have an affected child (**Mm**).

So why would more males than females be affected? What is the biological difference between men and women that might affect the manifestation of this disease? It is menstruation. Each month, women lose blood naturally through menstruation and so lose some iron in the process, whereas men obviously do not. This type of trait is called sex-influenced because, although the gene is not on a sex chromosome, the sex of the individual influences the severity of the phenotype.

Another Sex-Influenced Trait: Male Pattern Baldness

A more dramatic example of a sex-influenced trait is male pattern baldness. Because mostly men are affected, one might guess that it is a sex-linked trait. A classic case of male pattern baldness appears in President Adams's family. The second U.S. president, John Adams, and his son John Quincy Adams, the sixth president, as well as the latter's son and grandson, all had male pattern baldness. If this trait was on the X chromosome, as figure 3.2 shows, a father could not pass the trait onto his sons, meaning they must have received the trait from their mothers. However, it would be quite a coincidence if all four of the mothers of the individuals listed above were carriers of male pattern baldness. Actually, male pattern baldness is also a sex-influenced trait. The difference in the appearance of the trait is due to the hormonal differences between men and women. Thus in men the trait behaves as a dominant trait, while in women it behaves as a recessive trait.

Dominant Genetic Diseases

Although the majority of genetic diseases identified to date are recessive, there exist some dominant genetic diseases in humans. One of them is Marfan syndrome. This condition is due to a defect in the protein that forms connective tissues, such as tendons. The defect causes these connective tissues to be weaker; thus resulting in individuals that are tall and lanky and have long fingers and toes. These manifestations of the "disease" are not too troublesome. In fact, it may make afflicted individuals better basketball or volleyball players. Unfortunately, the defective protein is also necessary for the strength of the aorta, the major artery of the body. Because of a weaker aorta, some individuals with Marfan syndrome have died of sudden rupture of the aorta.

Those who know they are at risk for Marfan syndrome, because of their stature or family history, can get tested for the gene. Affected individuals can opt for preventive heart surgery that places an artificial tube to support the aorta. The Punnett squares in figures 3.3.C–D shows how to calculate the probability that an individual with Marfan will have an affected child. In the case of a dominant disease, the probability of having an affected child is 50 percent, even if only one of the parents is affected.

Huntington's disease is another dominant genetic disease. This condition does not affect individuals carrying the gene until well into adulthood. But then the condition slowly progresses over a five- to fifteen-year period, causing physical and mental deterioration due to nerve-cell death. Because this disease has different ages of onset and varying rates of progression, it is sometimes difficult to be certain of the diagnosis. People with this disease develop characteristic chorea, or convulsive, involuntary twitching movements. For this reason, this condition is also called Huntington's chorea. Here, also, the presence (or absence) of the gene can be tested for.

Pedigree Analysis

For all the above Mendelian traits in humans, we can use Punnett squares to determine the chances that one set of parents of a given genotype will produce offspring of a particular genotype and phenotype. For example, in the case of a recessive trait not linked to the X chromosome, two carrier parents have a 25 percent probability of producing homozygous, afflicted offspring (figure 3.3.B). These numbers are different if the trait is X-linked or dominant. But the key here is *probability*. Unlike the fruit fly or garden flowers, humans typically have only a few offspring. Under these circumstances, it is much more difficult to determine the rules of inheritance for a trait, even if one knows it is due to a single gene. To use another example, we have already seen that in humans an XX sex chromosome female mating with an XY sex-chromosome male, one expects to produce 50 percent sons and 50 percent daughters. Yet, how many of our families have exactly half boys and half girls? There is an element of chance, probability, which must be taken into account. In addition, we cannot just order two people with a particular genotype to produce ten children in order to determine the rules of inheritance! So how do we try to determine the rules of inheritance for a particular trait in humans?

Pedigree analysis is the tool used to solve this dilemma. Let us first learn the symbols used in establishing pedigrees (figure 3.4). By definition, circles represent females and squares represent males. A diamond is used if the sex of the individual is unknown. A horizontal line that connects two individuals indicates the mother and the father of a given offspring. The vertical line going down from the horizontal line joining a father and a mother indicates the lineage of their children. Individuals afflicted with a particular condition are represented

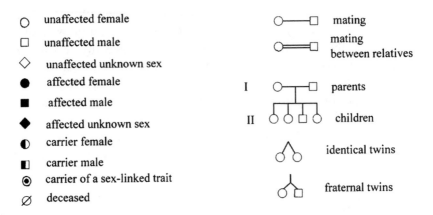

Figure 3.4 Standard Symbols Used in Pedigrees. Roman numerals indicate generations, and children are shown in their order of birth from left to right.

by filled symbols, while unaffected individuals are represented by open symbols. If it is known that an individual is a carrier for a non-sex-chromosome-linked trait, half-filled circles or squares are used. A carrier of a sex-linked trait is represented by a small dark dot inside the circle for that individual. A diagonal line through a symbol indicates a deceased individual. To facilitate discussion of the pedigree, roman numerals are used for successive generations. Individuals are also sometimes numbered for clarity. In large pedigrees, to avoid cluttering, individuals that do not influence the lineage are sometimes not shown.

Figure 3.5 shows Queen Victoria's pedigree for the hemophilia A trait. Notice that all the affected individuals are male. If one looks at Queen Victoria's ancestors, no one exhibited hemophilia. The defect responsible for the disease likely occurred in an X-carrying sperm from her father or in one of her mother's eggs, either of which was destined to become baby Victoria. Finally, unlike the case of male pattern baldness in President Adams's family, none of the sons in Queen Victoria's pedigree has an affected father.

Let us now consider an example of non-X-linked dominant trait described earlier, Marfan syndrome. We have already seen in figure 3.3.C–D how we can determine the chances that a parent with Marfan will pass on the trait to his or her offspring. A four-generation pedigree of a family with Marfan syndrome is shown in figure 3.6. Only

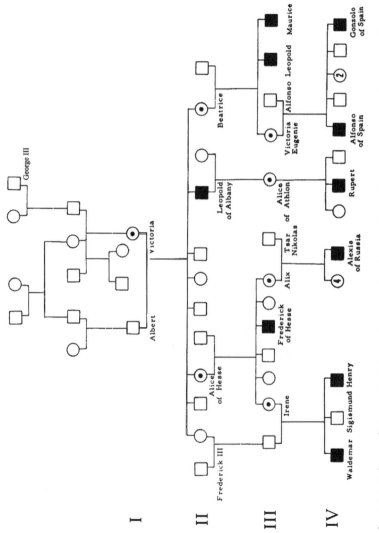

Figure 3.5 Pedigree of Hemophilia in Queen Victoria's Royal Descendants. Four generations are shown with affected males as filled in squares, and carrier (heterozygous) females shown as circles with a dark dot. Note that in this case, only males are affected, but no sons with hemophilia have fathers with hemophilia.

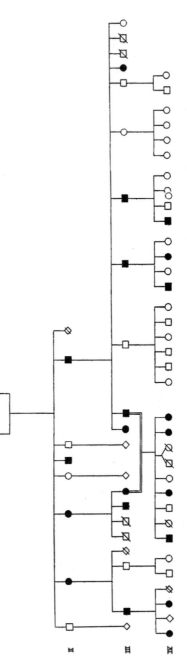

Figure 3.6 Pedigree of a Family with Marfan Syndrome. Because of the large number of individuals, spouses that married into the family are not shown. Note that there is an unbroken line from the father in the first generation to each of the affected great-grandchildren in the fourth generation.

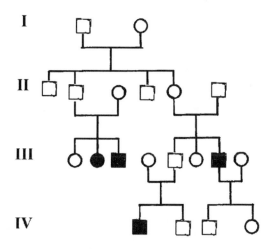

Figure 3.7 Pedigree of a Family with Alkaptonuria. Note that two sets of unaffected parents in the II generation and one set of parents in the III generation have children with alkaptonuria. That unaffected parents can have affected children is the hallmark of a recessive trait.

members of the family, not spouses that married into the family, are shown. We can trace directly back to the father in generation I from each of the nine affected individuals in generation IV. This is a hallmark for a dominant trait. Once the trait is lost in a family line, as seen in four branches of the family in this pedigree, it is completely lost from that branch.

Let us now study the pedigree of the recessive trait that causes the metabolic disease called alkaptonuria, in which an affected individual's urine turns dark and alkaline upon exposure to air. A family pedigree for alkaptonuria is shown in figure 3.7. The hallmark of a recessive trait is that parents who do not exhibit the trait can have children who do. To produce an affected offspring, the parents must both be heterozygous for the trait. By using a Punnett square, we can find out that there is a 25 percent probability that they will have affected children (figure 3.3.B). In the four generations of the family shown here, there are three sets of parents without alkaptonuria who have children with this trait. In one family, two out of three children are affected. That this proportion is much more than the 25 percent predicted from a Punnett square is merely due to random chance. This higher-than-expected proportion does not change the fact that alkaptonuria is a recessive trait.

Summary

We learned in this chapter that, as plants and fruit flies do, people also possess traits governed by single genes. These single-gene traits are determined by Mendelian rules of segregation and dominance. Some of these traits are associated with the X chromosome and are called sex-linked. We also learned that some traits that do not reside on the X chromosome are nevertheless affected by the sex of the person carrying that trait. These are called sex-influenced traits. The Punnett square allows us to estimate the chances of having children with a particular trait if we know the genotypes of the parents. Indeed, Punnett squares in some cases allow us to predict the genotype of the parents if we know the phenotype of their offspring and the rules of inheritance. However, in order to determine whether a trait in humans is dominant or recessive, we must analyze large pedigrees of affected families.

Try This at Home: Pedigree Game

Many people are interested in genealogy and their family history. In order to try to figure out the rules of inheritance, dominant or recessive, it is instructive to simulate a pedigree. The following exercise, though seemingly simple, accurately simulates how we inherit single-gene traits. You may find that trying to figure out whether something is dominant or recessive is not that simple. Remember this only applies to traits that are clearly dominant or recessive. Many traits of interest are not so clear cut.

We first start out with a genotype chart, a simple version of which is shown in figure B.3.1.A. In this simulation, as in real life, each parent contributes one of the two copies of the gene to each offspring. The choice of which version is transmitted is completely random. Thus to produce offspring in this simulation, flip a coin. Decide which copy of the gene corresponds to heads and which to tails, for example, heads = **A** and tails = **a**. Then determine the genotype of a select offspring. In this example, toss a coin once for the father. This determines which copy of the gene the father contributes to his offspring. You need not toss a coin

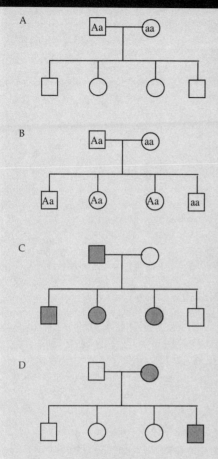

Figure B.3.1 Examples of Pedigree Charts. A. Genotype chart with just the genotype of parents shown. B. Genotype chart with the parents and potential offspring shown. C. Phenotype chart for the genotype chart shown in B for a dominant trait. D. Phenotype chart for the genotype chart shown in B for a recessive trait.

for any homozygous parent since both copies of his or her gene is the same. In the example given here, you need not toss the coin for the mother since she can only contribute an **a**.

Continue for each blank individual on the genotype pedigree chart. *Faithfully* note the result of each coin toss. Even if the expected ratio of offspring genotypes is 1 **Aa** : 1 **aa**, as in this example, if each time you toss the coin you came up with **Aa**, do not fudge the data, but write "**Aa**" for each offspring. By tossing a coin we are simulating nature; each offspring is a result of a random choice from the two copies of the gene in each parent.

After completing the genotype pedigree chart, fill in the phenotype pedigree charts appropriate for the trait that you are assigned. For example, let's say that we got the genotypes shown in figure B.3.1.B for the above chart.

If the trait were dominant, your phenotype chart would look like figure B.3.1.C

continued on next page

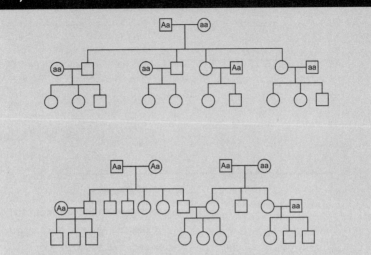

Figure B.3.2 Sample Genotype Pedigree Charts That Can Be Used for the Pedigree Game

On the other hand, if your trait were recessive, your phenotype chart would look like figure B.3.1.D

Study your phenotype chart. If you did not know originally that it was dominant or recessive, can you tell from the phenotype chart which one it was? Often, especially with a small number of family members, it is impossible to tell whether a trait is dominant or recessive. That is, you can write a genotype consistent with either hypothesis. There are, however, hallmarks for recessive and dominant traits, as discussed in chapter 3. Figure B.3.2 shows a couple of sample genotype charts that you can use for your simulation. Of course, you can make up any family pedigree chart to do this.

CHAPTER 4

From Genes to Phenotype

AS WE SAW IN CHAPTER 1, DNA is a very long, thin molecule. This molecule contains the genes that determine what an organism is and does. Since the only thing that varies among DNA molecules is the sequence of base pairs, not the sugar-phosphate backbone, the "program" encoded by DNA must reside in the order of base pairs. And indeed, it is helpful to think of DNA as an old-fashioned ribbon-like computer tape (now replaced by CDs) containing the software of the cell, this software being the base sequence. As in any computer, the software can only be executed if the proper hardware decodes the instructions present in that software. This is in fact how the genes present in DNA carry out their commands: the language of DNA base pairs is interpreted and executed by the cell's hardware. The decoding of DNA takes place in two separate steps, called transcription and translation, respectively.

Proteins are the end products of the flow of genetic information in living cells. They directly determine how an organism looks and acts. The function of transcription is to produce a ribonucleic acid (RNA) copy of the DNA, while the function of translation is to decode this RNA copy and direct the production of proteins. Proteins can be thought of as agents that determine the phenotype of cells and organisms, while the genotype is stored in DNA.

RNA   # DNA

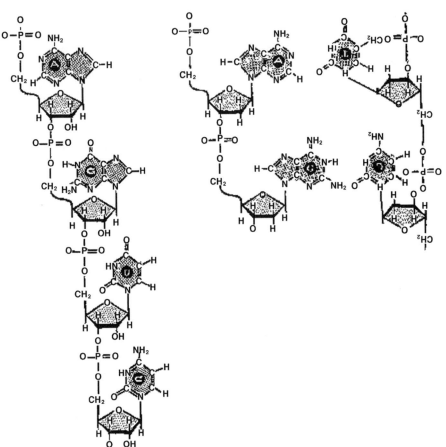

Figure 4.1 The Structures of DNA and RNA Contrasted. DNA is a double-stranded molecule with deoxyribose as the sugar, while RNA is a single-stranded molecule with ribose as the sugar. The thymine in DNA is replaced by a similar structure, uracil, in RNA.

Transcription

In all cells, the genetic message encoded in the DNA is first transcribed into a faithful RNA copy of the DNA molecule. The two key differences between DNA and RNA molecules are that the sugar in RNA is ribose instead of deoxyribose, and the base uracil replaces thymine (figure 4.1). In addition, RNA molecules are not double helices like DNA. We say that RNA is a single-stranded molecule, not a

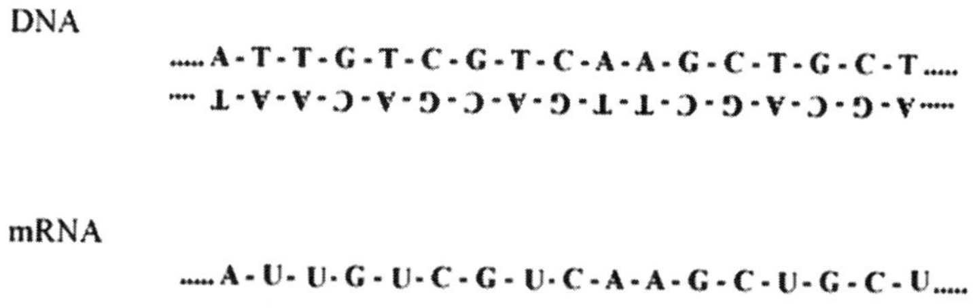

DNA

.....A - T - T - G - T - C - G - T - C - A - A - G - C - T - G - C - T .....

···· T · A · A · C · A · G · C · A · G · T · T · C · G · A · C · G · A ·····

mRNA

.....A - U - U - G - U - C - G - U - C - A - A - G - C - U - G - C - U.....

Figure 4.2 DNA and its mRNA Copy. A short piece of double-stranded DNA, represented by letters, and its mRNA copy.

double helix (or a double-stranded molecule) like DNA. However, the base sequence found in messenger RNA (mRNA), as RNA copies of DNA are known, has the exact same base sequence as that found in DNA, with U replacing T (figure 4.2).

The hardware that copies genes present in the DNA is called RNA polymerase. This enzyme reads the base sequence of DNA and builds an RNA copy of each gene in the DNA molecule. To do this, RNA polymerase binds tightly to DNA, moves along it, and makes an RNA copy using RNA building blocks in the cell. This process is like copying DNA, except that only one of the two DNA strands is copied, thus the RNA product is released as a single-stranded RNA molecule (figure 4.3). Also, each gene is copied separately by RNA polymerase. For this, the base sequence of each gene has at the beginning what is called a "promoter region." This is the part of the DNA that RNA polymerase recognizes as the "start" signal for a gene. Think about a promoter as being a DNA base-pair sequence that tells RNA polymerase to "begin here." Transcription of the gene can then proceed. When RNA polymerase reaches the end of the gene, a terminator region, also made of a special base sequence, tells RNA polymerase to "stop transcribing here." The RNA produced is called messenger RNA (or mRNA). RNA polymerase then falls off the DNA and is ready to perform another round of transcription (figure 4.4).

Thus, a cell contains a large collection of different messenger RNA molecules, each a copy of a particular gene. Bacteria typically harbor a few thousand genes, while humans have approximately 35,000. This means that bacterial cells can contain a few thousand different messenger RNA molecules, while human cells can contain tens of thousands. There may be just a few messenger RNA for one gene and

Figure 4.3 The Beginning of Transcription. A diagram illustrating how transcription starts with the DNA double strand opened up and single-stranded mRNA copied from one of the DNA strands. The lighter-shaded line represents the sugar-phosphate backbones of the DNA and RNA. The rectangular blocks hanging from these backbones represent the bases. The bases of mRNA are complementary to the DNA bases from which they are being copied.

hundreds of messenger RNAs for another gene. The lengths of messenger RNA molecules also vary from about 1,000 bases in bacteria to tens of thousands of bases in humans. This is because bacterial genes are in general much shorter than human genes.

Translation

Messenger RNA molecules are an intermediate step in the decoding of genes; they must next be translated into protein products. The machinery that translates mRNAs is quite complicated, and it took many years to unravel the mystery of their functioning. Proteins are polymers of twenty different amino acids arranged in a particular sequence. This is true of proteins from viruses to human beings. Hundreds of thousands of different proteins exist in the living world. They differ by the arrangement and length of amino acids (table 4.1). The central question then is this: how does the translation machinery "know" where to insert a given amino acid in a growing protein molecule and how does it "know" where to start and where to stop the synthesis of a particular protein? This is done by the cellular decoding machinery that translates the sequence of nucleotides in mRNA into a sequence of amino acids.

Each amino acid is coded for in DNA by a sequence of three adjacent base pairs. A set of three contiguous base pairs is called a "codon." For example, ATG determines the amino acid methionine. Most amino acids are coded for by several codons: for example, both AAA and AAG code for lysine. Some amino acids are coded for by as many as six different codons. Interestingly, three codons, TAA, TAG, and TGA do not correspond to any amino acids; they mean "stop" and instruct the translation machinery to stop making a protein molecule.

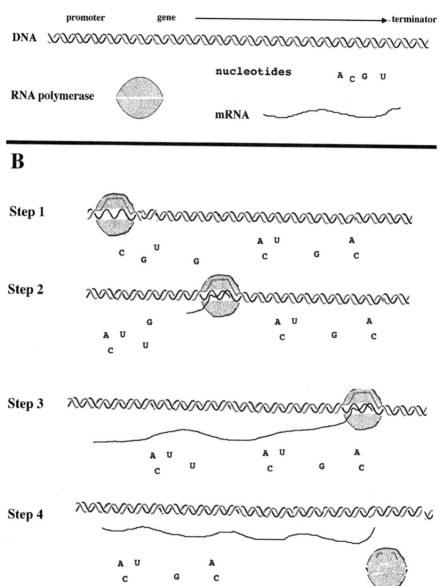

Figure 4.4 Transcription. A. The gene in the DNA has a region known as the promoter that signals the RNA polymerase to bind to the DNA and begin making RNA using nucleotide subunits. DNA is diagrammed as a double helix, with the strand that is read printed darker than the opposite strand. The gene also has a terminator region that signals the end of the gene. B. The process of transcription. In Step 1, RNA polymerase binds to the promoter. Step 2 shows the initiation of mRNA production. Step 3 shows a long piece of mRNA, attached to the RNA polymerase, towards the end of the process. Transcription ends in Step 4 when the RNA polymerase finds the terminator and falls off the DNA.

Table 4.1 Examples of Proteins a   nd Their Amino Acid Compositions

Amino acid	(abbrev)	Hemoglobin	DNA polymerase	Insulin	Fibrillin	Rubisco
Alanine	(A)	19	91	1	96	44
Arginine	(R)	8	76	1	131	29
Asparagine	(N)	1	12	3	188	16
Aspartic acid	(D)	8	42	0	173	26
Cysteine	(C)	2	0	6	361	9
Glutamine	(Q)	5	16	3	101	11
Glutamic acid	(E)	5	87	4	201	32
Glycine	(G)	6	58	4	307	46
Histidine	(H)	8	18	2	48	14
Isoleucine	(I)	1	25	2	148	22
Leucine	(L)	25	124	6	141	40
Lysine	(K)	5	42	1	111	23
Methionine	(M)	0	16	0	52	9
Phenylalanine	(F)	6	27	3	84	22
Proline	(P)	6	48	1	175	20
Serine	(S)	15	31	3	173	17
Threonine	(T)	6	30	3	166	28
Tryptophan	(W)	1	14	0	13	8
Tyrosine	(Y)	4	24	4	94	17
Valine	(V)	10	51	4	18	32
total #aa		141	832	51	2781	465

Compositions are for:

Human β-hemoglobin, the protein responsible for carrying oxygen in our blood. This is the protein that is mutated in sickle-cell anemia.

DNA polymerase from *Thermus aquaticus* used in PCR reactions.

Human insulin, the protein necessary for regulating blood sugar.

Human fibrillin, a connective tissue protein, which if mutated can cause Marfan's syndrome.

Rubisco (ribulose bisphosphate carboxylase), which is the most abundant protein in the world and necessary for all of life. It is present in all plants and is used to convert the energy from the sun to fix carbon; that is, to make carbon available to nonplants. This is the composition of rubisco from sugar maple.

You may already have guessed that stop codons are found at the end of a gene. In addition, there is a single "start" codon, ATG, that, we just saw, codes for methionine. Does this mean that all proteins start with the amino acid methionine? The answer is yes, with very few exceptions. However, many proteins are processed further after they are made, and, sometimes, this process involves removing some amino acids at the beginning of the protein. This is the case for both hemoglobin and insulin listed in table 4.1.

A.

DNA

.....ATG AGG GGG CTG CCA TGT CCC CAG AGA GCA GTG TTG GGC CTG AGC ATG GCA GGG TGA..

¨¨¨TCA TCC CCC GAC GGT ACA GGG GTC TCT CGT CAC AAC CCG GAC TCG TAC CGT CCC ACT¨

B.

Corresponding mRNA

AUGAGGGGGCUGCCAUGUCCCCAGAGAGCAGUGUUGGGCCUGAGCAUGGCAGGGUGA

C.

Corresponding protein (using single-letter abbreviations)

M-R-G-L-P-C-P-Q-R-A-V-L-G-L-S-M-A-G

Figure 4.5 Converting a Gene to a Phenotype. A. A short stretch of double-stranded DNA. B. The corresponding mRNA. C. The corresponding protein made using this gene.

Figure 4.5 shows a hypothetical very short gene (the DNA double helix), its mRNA copy (where U replaces T), and the protein product of that gene. The gene has an ATG codon at the beginning, and one of the stop codons (TGA) at the end. Note that all the codons (except TGA) direct that a particular amino acid be placed in a particular position in the protein. The codes for all twenty amino acids and the "start" and "stop" codons are known (figure 4.6), and these are called the genetic code. The genetic code is universal (with very minor exceptions), a fact that led James Watson to claim that "what's true for *E. coli* [a bacterium] is true for an elephant."

At this point, let us examine in greater detail the process of translation. To understand this mechanism, it is useful to think in terms of two different languages. In that sense, both DNA and RNA "speak" a language where the "words" are codons made up of three bases. RNA "speaks" almost the same language as DNA; let us call it a "dialect" where U replaces T. The DNA language and the RNA dialect are in fact indistinguishable because U and T have the same "meaning." However, proteins do not "speak" the base language or dialect: they speak the language of amino acids. How is one language (and its dialect) translated into the other? In human affairs, an interpreter does the job. Does it work the same way in living cells? Yes, it does.

First position	Second position				Third position
	U	C	A	G	
U	UUU Phe ⎫F UUC Phe ⎭ UUA Leu ⎫L UUG Leu ⎭	UCU Ser ⎫ UCC Ser ⎬S UCA Ser ⎪ UCG Ser ⎭	UAU Tyr ⎫Y UAC Tyr ⎭ UAA Stop UAG Stop	UGU Cys ⎫C UGC Cys ⎭ UGA Stop UGG Trp **W**	U C A G
C	CUU Leu ⎫ CUC Leu ⎬L CUA Leu ⎪ CUG Leu ⎭	CCU Ser ⎫ CCC Ser ⎬P CCA Ser ⎪ CCG Ser ⎭	CAU His ⎫H CAC His ⎭ CAA Gln ⎫Q CAG Gln ⎭	CGU Arg ⎫ CGC Arg ⎬R CGA Arg ⎪ CGG Arg ⎭	U C A G
A	AUU Ile ⎫ AUC Ile ⎬I AUA Ile ⎪ AUG Met **M**	ACU Thr ⎫ ACC Thr ⎬T ACA Thr ⎪ ACG Thr ⎭	AAU Asp ⎫N AAC Asp ⎭ AAA Lys ⎫K AAG Lys ⎭	AGU Ser ⎫S AGC Ser ⎭ AGA Arg ⎫R AGG Arg ⎭	U C A G
G	GUU Val ⎫ GUC Val ⎬V GUA Val ⎪ GUG Val ⎭	GCU Ala ⎫ GCC Ala ⎬A GCA Ala ⎪ GCG Ala ⎭	GAU Asp ⎫D GAC Asp ⎭ GAA Glu ⎫E GAG Glu ⎭	GGU Gly ⎫ GGC Gly ⎬G GGA Gly ⎪ GGG Gly ⎭	U C A G

Figure 4.6 The Genetic Code. Note that amino acids can be coded for by several codons. Special codons for start and stop are shown in dark boxes.

In the early 1960s, researchers discovered a class of small RNA molecules, called transfer RNAs (tRNAs), that have the ability to bind specific amino acids. What's more, it was realized that these tRNAs also had the ability to "recognize" mRNA codons via three complementary bases that they possess in what is called an anticodon. Therefore, these tRNAs can communicate with both the "base language" (thanks to their anticodons) and the "protein language" (thanks to their ability to bind amino acids). What happens (figure 4.7) is that tRNAs loaded with their amino acids line up along an mRNA molecule, interact through their anticodons with specific mRNA codons, and, by doing so, bring the amino acids they carry into very close contact. These amino acids can then form chemical bonds between one another and so produce a protein.

The "assembly line" containing the mRNA, tRNAs, and a growing protein needs to be held together in a precise fashion to function

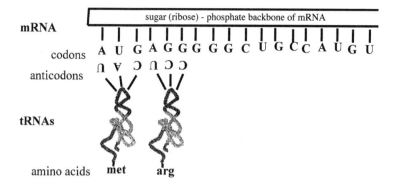

mRNA

Figure 4.7 Translation. Messenger RNA encodes a gene depicted at the top in triplet codons. Transfer RNAs (tRNAs), which translate the triplet codes into a sequence of amino acids, are depicted by the curlicue structure of their sugar-phosphate backbones. At one end of the tRNA are three letters corresponding to the triplet anticodons that are complementary to the codons in mRNA. The corresponding amino acids are attached to the opposite ends of the tRNAs. The process of making proteins involves breaking the bonds between amino acids and tRNA and making new bonds to hold the amino acids together in the new protein.

properly. This is achieved in cells through interactions with particles called ribosomes. Ribosomes are composed of many proteins and several RNA molecules. These are arranged in such a way that ribosomes can slide along an mRNA molecule, while cavities in the ribosome structure hold tRNA molecules loaded with an amino acid in the right orientation. This ensures that codons, anticodons, and amino acids are aligned properly in close proximity and in the right order. The full protein "assembly line" is depicted in figure 4.8. Note that protein synthesis start at the AUG RNA codon (with methionine). At a UGA codon, the ribosome cavity facing a UGA codon remains empty because no tRNA with an ACU anticodon exists, thus protein synthesis stops. The now fully formed protein then leaves the ribosome and becomes free to exercise its function where it is needed.

Transcription and translation basically work the same way in all living cells, from bacteria to humans.

Changes in DNA Modify the Amino Acid Sequences of Proteins

Given that codons in the DNA completely determine the nature and location of amino acids in proteins, it is easy to see that if, for any reason, codons in a gene change, this will result in the insertion of a different

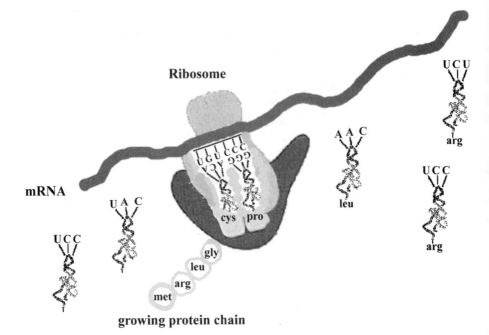

Figure 4.8 A Protein Being Made from mRNA. A cartoon image of protein being made off of mRNA. The wavy line is the mRNA with a ribosome in the process of translating the sequence of triplet codons into their corresponding amino acids. Once the amino acids on the tRNA are attached to the growing protein chain, tRNAs without amino acids (on the left) are released. Then the ribosome moves over to the next codon and another tRNA with the complementary anticodon with its amino acid comes into the ribosome.

amino acid in a protein. This is basically how mutations (changes in DNA base pairs) manifest their effects. Mutations will be studied in chapters 7 and 8. For now, let us just look at a few consequences of some base pair changes in the gene that codes for β-hemoglobin, one of the protein chains that constitute hemoglobin. We saw in chapter 3 that sickle-cell anemia is a very serious genetic disease. Its molecular basis is very well understood. Let us look at a small portion of the normal β-hemoglobin mRNA, which corresponds to codons 5–8:

... CCA-GAA-GAG-AAG. ...

The genetic code indicates that the translation product of this mRNA is the amino acid sequence

... Proline–Glutamic acid–Glutamic acid–Lysine. ...

It turns out that this gene is mutated in sickle-cell anemia patients and produces a mRNA with the sequence

... CCA-GUA-GAG-AAG. ...

You will notice that the second base of the second codon (in bold) is now a U instead of an A. This change encodes a protein with the amino acid sequence

... Proline–Valine–Glutamic acid–Lysine. ...

Thus, the amino acid valine replaces the glutamic acid found in normal hemoglobin, because GUG codes for valine while GAG codes for glutamic acid. This single amino-acid substitution occurred because the β-hemoglobin gene underwent a mutation, whereby an A-T base pair in the DNA was turned into a T-A base pair. In addition, this single change of an amino acid is responsible for all the problems associated with sickle-cell disease. This is because mutant β-hemoglobin, with this seemingly minor change from a glutamic acid to a valine, has a strong tendency to crystallize in red blood cells and sickle them, contrary to normal β-hemoglobin. Many genetic diseases are the result of such subtle mutational changes in codons.

Gene Regulation

The genome is the sum total of all the genes present in an organism. For the organism to function properly, not all genes in the genome can be active at all times. Similarly, not all cells of a multicellular organism express all the genes present in its genome. For example, human pancreatic cells do not make hemoglobin, but they do produce insulin. On the other hand, progenitors of red blood cells synthesize hemoglobin but do not make insulin. Also, human embryos produce a special hemoglobin that is not found in the adult. How is this possible, knowing that all cells in the human body contain the same DNA? The answer is that cells can turn genes on and off to regulate which genes are expressed in what cells and when. We say that the insulin gene is turned off in all cells except specialized pancreatic cells, whereas the gene coding for embryonic hemoglobin is turned on during fetal life but is turned off later on.

Gene regulation is very complicated and only partially understood. It takes place in all living organisms, from bacteria to humans. Often, whether a gene is active (transcribed) or inactive (not

A

promoter   gene   ──────────────────────────▶ ·terminator

DNA

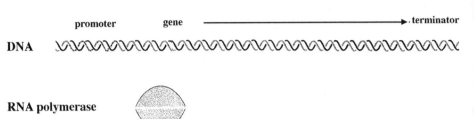

RNA polymerase

B
Promoter not blocked, so transcription can occur

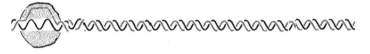

C
Protein factor bound to the promoter to prevent transcription

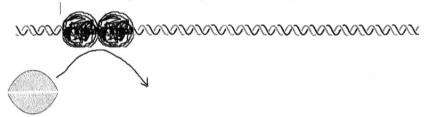

Figure 4.9 Regulation of Transcription. A. Definitions of symbols used in this figure. B. A portion of DNA diagrammed to show that when RNA polymerase binds the promoter region, transcription can proceed. C. If a protein factor binds to the promoter region of the DNA, RNA polymerase cannot bind and thus the protein factor prevents transcription of the gene.

transcribed) depends on interactions between the gene's promoter and protein factors present in a given type of cells. Basically, when these protein factors are present, they bind to the promoter region and physically block it (figure 4.9). Under these circumstances, RNA polymerase has no access to the promoter and hence cannot transcribe the gene. Without mRNA production, the gene cannot produce its protein, and thus the gene is turned off. Conversely, other factors can bind to this blocking protein factor and *prevent* it from blocking the promoter. In this case, RNA polymerase has access to the promoter. The gene is then transcribed, its mRNA subsequently translated, and the gene is turned on. Cells contain thousands of such factors that act in concert to fine-tune gene activity.

Embryonic Stem Cells Exhibit a Special Type of Gene Regulation
Embryonic stem cells provide an interesting example of gene regulation. Embryos contain stem cells, cells that can be coaxed into producing any and all types of human tissue. By definition, a young embryo contains cells that will eventually produce all the organs found in a complete human being. In other words, the cells of a young embryo are pluripotent, and can differentiate into any cell type, such as brain, heart, or muscle cells, and so forth. Researchers are keen on understanding what triggers embryonic stem cells to differentiate into specialized tissues. Once these mechanisms are understood, it will be possible to regenerate organs that are faltering due to accident or illness. The idea of repairing the spinal cord of paraplegic and quadriplegic individuals comes to mind. For now, suffice it to say that not much is known about the signals that allow embryonic stem cells to differentiate into various organs. One thing is clear, however. Before differentiation, embryonic stem cells express a large collection of genes; many of their genes are "on." As the cells differentiate, many genes are progressively turned "off," and specialized genes, which are only expressed in differentiated organs, are turned "on."

Box 4.1 *Why People Are Saving Their Babies' Cord Blood*

Because stem cells are capable of turning into a variety of different specialized cells, such as brain cells or blood cells, they hold great promise for the treatment of diseases such as Alzheimer's, heart disease, and a variety of cancers. Cells in an early embryo, just by their very nature, can give rise to all the cells in the body. However, embryonic stem cells are controversial because their use necessitates killing the embryo.

Another source of stem cells is the umbilical cord blood. These stems cells can give rise to cells in the blood including red blood

continued on next page

Box 4.1 *continued*

cells, which carry oxygen, white blood cells, which are cells of the immune system, and platelets, which are involved in blood clotting. Similar stem cells are found in adult bone marrow. But the stem cells in the cord blood are younger and thus have a higher chance of providing a good match. Cord blood also has a higher number of stem cells. Until recently, the blood in the umbilical cord was routinely discarded with the placenta. We now can collect and store the stem cells from cord blood. The procedure requires that cord blood be collected soon after the umbilical cord is clamped off. Special processing and freezer storage makes the stem cells available years later to the individual from which the cord blood was collected. Although not all diseases treatable by stem cells can be treated with cord-blood stem cells, a large number of diseases associated with blood, immune systems, and other metabolic diseases can be treated with cord-blood stem cells.

In the United States, cord blood may be donated to many public blood banks, including the American Red Cross, for use by any one. The National Institutes of Health also maintain a cord-blood bank. Donating your baby's blood to these two cord-blood banks does not cost the donor. Alternatively one can pay approximately $2,000 to store the baby's cord blood in a private cord bank, Cord Blood Registry (http://www.cordblood.com), and have it available for the donor's use.

One report of how cord blood can save a life was reported on CBS's *60 Minutes II*. This is the story of Keone Penn, a teenager with sickle-cell anemia. This is a disease of hemoglobin in the red blood cells. In the past, the only possibility for a cure was a bone marrow transplant. However, for Keone, there was not a good donor match. But doctors thought that cord blood might help to cure Keone of his sickle-cell disease. It would be easier to find a good match of cord blood, and cord blood has higher numbers of stem cells than even bone marrow. Fortunately for Keone, a good match was found in the New York Public Blood Bank. His own bone marrow stem cells with the sickle-cell gene were killed with chemotherapy, then he received the matching cord blood, which

Box 4.1 *continued*

proceeded to make all the different blood-cell types, including normal red blood cells without sickle cells.

Leukemia is a cancer of the blood cells. Cord blood can be used to replace the cancerous cells later in the life of the donor, but in an unusual twist, cord blood can even save the life of the donor's mother! Patrizia Durante was pregnant with her first daughter when she was diagnosed with leukemia. Her daughter was delivered premature, and Patrizia was aggressively treated with chemotherapy to rid her of her leukemia. Again, the traditional treatment is to look for matching bone marrow, but again her doctors could not find a match. So they used her baby's cord blood instead. Not only did the baby's stem cells start rebuilding her mother's blood, they also attacked the cancerous cells.

With these and many other successes, there is increasing interest in cord-blood banking. We can now make good use of blood that until recently was thrown into the garbage.

Summary

The genetic information, stored in DNA base pairs, is first converted into messenger RNA molecules (mRNA) that are copies of the DNA genes. This process is called transcription. Regulation of transcription is performed by base sequences that represent transcription start and stop sites, called promoters and terminators. The messenger RNA molecules are then translated into the amino acid sequence of proteins by transfer RNAs (tRNAs) and ribosomes. The genetic code consists of codons, three-base-pair sequences that specify amino acids, as well as translation start and stop codons necessary to form proteins. As a result, changes in codons at the DNA level, mutations, can result in the incorporation of the incorrect amino acids into proteins. Proteins with altered amino acid sequences can malfunction and lead to a diseased phenotype. Finally, we also saw that gene activity can be turned on and off at the level of transcription by protein factors that can prevent or allow transcription.

With this game, we can graphically see how genes are copied, transcribed, and translated into protein. The purpose of this game is to faithfully replicate a piece of DNA and translate it into the correct protein sequence. It is written as a game involving a number of teams, but if you want to do this on your own, just don't check your sequence with the earlier sequence.

Procedure

The first member of each team will be given a DNA sequence, which may not be shown to the other members. The team members verbally "replicate" the DNA five times. For example, if person #1 has the sequence ATG, #1 figures out the complementary sequence, TAC, and says to #2, "TAC." Person #2 writes TAC, figures out the complementary sequence, ATG, and says "ATG" to #3, and so forth until person #5. Person #5 writes down the sequence he or she gets from #4 then determines the mRNA complementary to it. In this case, it would be UAC, so person #5 says "UAC" to #6. Then #6 determines the amino acid sequence corresponding to the mRNA using the single letter amino acid abbreviation given in figure 4.6. In this case, it would be the amino acid tyrosine, or Y. Team members may help in translation but may *not* provide the DNA sequence they wrote down.

The first team finished wins if the replication, transcription, and translation are 100 percent correct. If there are any errors in replication or translation, they will lose 10 seconds for each error in replication that does *not* result in an error in the final protein sequence and 20 seconds for each error in protein sequence. Then the resulting adjusted time determines the winner. Often in such a game, mistakes are made in the process. The longer the sequence of DNA, and more times that it is copied, the higher the chance for errors in the sequence, that is, mutations. This game imitates real life well because larger genes have a higher chance of mutating, and also, the more often genes are copied, the higher the chances of mutations.

Here are a couple DNA sequences that make an interesting translated amino acid sequence expressed by the single letter abbreviation of amino acids.

CCTCTTTTGCTCTGATAAACATCGTATTCGTTGCTTCGCTGTATT
ACCCTTGTGCGGCAACTCCCTGTTGTCCTACCCCTCTTACTCTCG

CHAPTER 5

Using Bacteria as Protein Factories

WE SAW IN CHAPTER 4 that the genetic code is universal; that is, codons are the same in all living organisms. This means that genes from one type of organism, animal cells for example, should be able to guide the synthesis of the corresponding protein in bacterial cells. In order to accomplish this, genes from one organism must be transferred to bacteria. We already saw in chapter 1 that DNA from one strain of bacteria can be transferred to another. This process is called bacterial transformation. Since DNA from different organisms has the same basic structure, only differing in the sequence of bases, we should be able to transfer DNA from any organism into bacteria. Given this ability, we should be able to manufacture animal, or even human, proteins in bacteria. This feat has indeed been achieved and is described in this chapter. Today, many important human proteins are made in bacteria, and these proteins are actually used in human medicine.

Tools for Manufacturing Proteins

Many proteins are medically useful. How would you go about making a lot of a medically important human protein? Isolating such proteins from humans can be problematic. For example, many proteins can only be found in substantial amounts in vital organs. Thus quantities are limited to those that can be obtained from the cadavers of people willing to donate their organs. Some proteins can be isolated from blood. However, blood can be contaminated with viruses such as HIV, hepatitis, and, more recently, West Nile virus. Expensive screening is required to ensure safety. You can see that protein availability is limited from these sources. Thus making the desirable proteins in bacteria is an attractive option. This involves a technique known as recombinant DNA technology, or cloning, and makes possible the production of human proteins in bacteria. The DNA technology that is used to do this relies on two critically important elements: restriction enzymes and plasmid vectors. Let us start with restriction enzymes.

These enzymes, also called restriction endonucleases, exist in all bacterial cells and have one specific function: they cut the sugar-phosphate backbones of "foreign" DNA, such as the DNA from viruses that infect and kill bacteria. Thus, restriction enzymes can be seen as defense mechanisms evolved by bacteria to defend themselves against viral attacks. Indeed, when restriction enzymes cut DNA, they break up the genes located on this piece of DNA. However, restriction enzymes do not cut DNA randomly, they only cut it at a *specific sequence*. For example, one restriction enzyme from the bacterium *Escherichia coli* (*E. coli*) cuts DNA when it encounters the sequence shown in in figure 5.1.A.

Another restriction enzyme from the bacterium *Haemophilus influenza* cuts it at a difference sequence, as shown in figure 5.1.B.

Note that the two sites where the restriction enzymes cut have an interesting structure. If one reads the nucleotides on one strand, the opposite strand read backward is identical. This is called a palindrome. A great majority of restriction enzymes cut at palindromic sequences. Restriction enzymes cut at these palindromic sequences of DNA double helix in a very special manner: they produce DNA fragments with complementary single stranded sequences at their ends. One can think of them as "sticky" ends of the DNA. That is, if another piece of DNA is cut with the same restriction enzyme, the two cut ends will have short complementary pieces that can base pair together. You will see below how scientists have taken advantage of this property. Over a hundred bacterial restriction enzymes in pure form are currently available commercially. This means that when scientists isolate DNA from any source, they can cut it with any of the available restriction enzymes. The DNA molecules obtained after cutting are much shorter and therefore much more manageable than the original

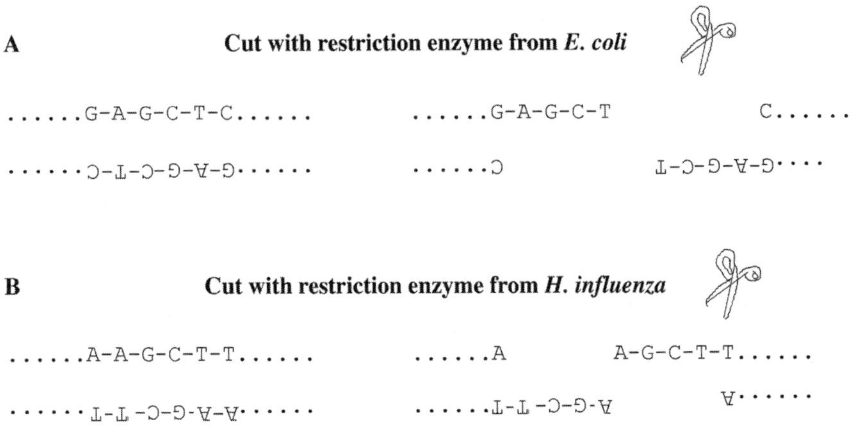

A   **Cut with restriction enzyme from *E. coli***

......G-A-G-C-T-C......   ......G-A-G-C-T   C......

······Ɔ-Ʇ-Ɔ-Ә-∀-Ә······   ······Ɔ   Ʇ-Ɔ-Ә-∀-Ә····

B   **Cut with restriction enzyme from *H. influenza***

......A-A-G-C-T-T......   ......A   A-G-C-T-T......

······Ʇ-Ʇ-Ɔ-Ә-∀-∀······   ......Ʇ-Ʇ-Ɔ-Ә-∀   ∀······

Figure 5.1 Examples of Restriction Enzyme Sites. The sites are shown whole on the left, and on the right they are shown after being cut by the enzyme. A. Restriction enzyme site for a restriction enzyme from *Escherichia coli*. B. Restriction enzyme site for a restriction enzyme from *Haemophilus influenza*.

very long stretches of DNA found in cells. Also, since a given restriction enzyme always cuts DNA at the same sequence, *these DNA fragments are a collection of pieces that begin and end with the same base sequence.* In addition, these short DNA fragments each contain a limited number of genes. Let us leave restriction enzymes for the time being and focus now on the second ingredient of cloning experiments: plasmid vectors.

Back in the 1960s and 1970s, scientists discovered that bacterial cells often contained more than one piece of DNA. The majority of the DNA was associated with the bacterial chromosome, a single piece of DNA containing several million base pairs and carrying several thousand genes. However, many bacterial species also harbored a circular "minichromosome" that only contained a few thousand base pairs and carried just a few genes. These "minichromosomes" are called "plasmids." An important property of plasmids is that a bacterial cell can often carry multiple copies, sometimes hundreds, of these "minichromosomes." Another important feature is that, being small, plasmids often contain just a single cutting site for a number of restriction enzymes. Thus, cutting a circular plasmid with a restriction enzyme generates a linear molecule with complementary single-stranded ends (figure 5.2).

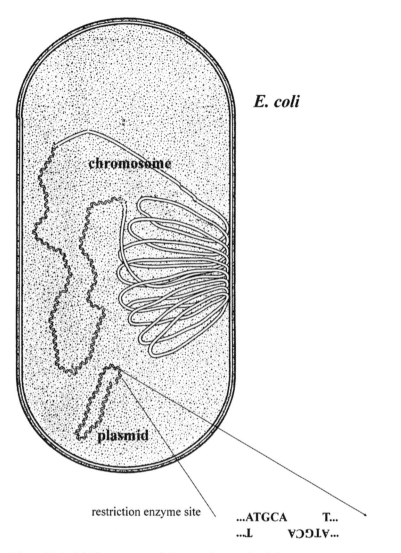

E. coli

chromosome

plasmid

restriction enzyme site

...ATGCA   T...
...ꓕ   ⱯƆ⅁ꓕⱯ...

Figure 5.2 *E. Coli* Chromosomes. A diagram of an *E. coli* cell showing its main chromosome and a minichromosome or plasmid. Because plasmids are much smaller, a restriction enzyme often cuts it at only a single site.

Using Restriction Enzymes and Plasmids to Clone a Gene

We will use the example of the insulin gene to explain how gene cloning works (figure 5.3). Insulin was the first protein to have its amino acid sequence deciphered. This is because it is relatively small (see table 4.1). It is also one that is very important medically. Figure 5.3.A shows a part of the amino acid sequence of human insulin. Because we know the genetic code if we know the sequence of amino acids of a protein, we can make a sequence of DNA, a gene, that corresponds to the sequence of amino acids in that protein. Figure 5.3.B shows the sequence of DNA corresponding to the amino acid sequence shown in figure 5.3.A. Because many amino acids can be coded for by a number of different codons, the DNA sequence shown is one of many that corresponds to the amino acid sequence. The gene coding for insulin was synthesized in this way in the laboratory. In order to have this gene transcribed at a high level in bacteria, a promoter and terminator are added to the beginning and end of the gene (figure 5.3.C). Finally, in order to clone the piece of DNA with the insulin gene on it, single-stranded ends complementary to the ends of plasmid DNA cut with a restriction enzyme are attached (figure 5.3.D). Next, scientists grow a culture of bacterial cells that contain plasmids. This plasmid DNA is purified and cut with the restriction enzyme that will produce the same complementary or "sticky" ends that were put on the piece of the insulin gene (figure 5.3.E). The engineered insulin gene is then mixed with the cut plasmid DNA. Because the cut plasmid molecules and the engineered insulin gene possess the same complementary sequences at their ends, these two DNA pieces will stick together (figure 5.3.F). Finally, the sugarphosphate backbones of the two DNA molecules are joined together with an enzyme known as DNA ligase. The resultant piece of DNA is referred to as a "recombinant DNA molecule" because two pieces of DNA from different sources are combined.

A process similar to making a gene from a known amino acid sequence, the product of which we shall refer to as "synthetic genes," is used for other relatively small proteins of medical interest. It is also used to make the Bt toxin gene used in genetically modified plants, which will be discussed in chapter 6. However, other proteins of medical interest are far too large to use this procedure. For those proteins, a more arduous and complex cloning procedure is necessary.

Producing Human Proteins in *E. Coli*

Once the above operations are finished, it is necessary to make many copies of the pieces of recombinant DNA. Copying uses the natural copying mechanism of live bacteria by introducing the recombinant DNA molecule that was made in the test tube into bacterial cells, generally *Escherichia coli* (*E. coli*). When treated with appropriate chemicals,

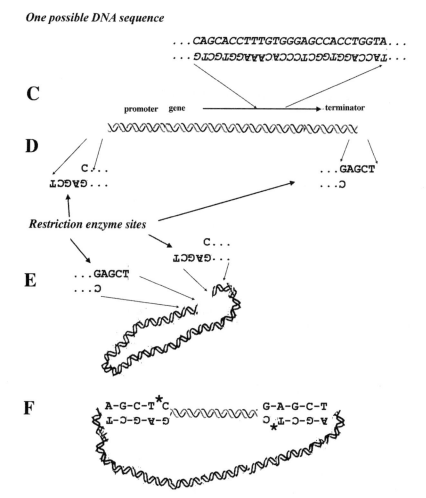

A ...Q-H-L-C-G-S-H-L-V...

B

One possible DNA sequence

...CAGCACCTTTGTGGGAGCCACCTGGTA...

...GTCGTGGAAACACCCTCGGTGGACCAT...

C

promoter   gene ———————→ terminator

D

C...   ...GAGCT

GAGCT...   ...C

Restriction enzyme sites

C...

GAGCT...

E

...GAGCT

...C

F

A-G-C-T*C   G-A-G-C-T

G-C-G-A-G   C*A-G-C-T

Figure 5.3 Making Human Insulin Using Recombinant DNA. A. Partial amino acid sequence of human insulin. B. The DNA sequence corresponding to the amino acid sequence shown in A. C. Promoter and terminator suitable for high-level transcription in *E. coli* are added. D. The insulin gene with the promoter and terminator is flanked by restriction enzyme sites that correspond to the restriction enzyme used to cut the plasmid. E. Plasmid DNA is cut with the restriction enzyme. F. Pieces of DNA with the insulin gene and the cut plasmid DNA join spontaneously at their complementary ends. Finally, an enzyme is used to glue the sugar-phosphate backbones of the two DNA pieces. The * indicates where the ligase enzyme glues the sugar phosphate backbone on the recombinant DNA. This process results in a recombinant DNA molecule containing the human-insulin gene in a bacterial-plasmid DNA. The engineered human-insulin gene is shown as the straight, short double helix, and plasmid DNA is shown as a curved double helix.

E. coli cells can take up DNA molecules in their growth medium. Once inside the bacterial cells, the recombinant DNA molecules can be copied, just like plasmids naturally found in *E. coli*. In fact, the transformed *E. coli* cells do not "know" that the plasmids now contain human DNA, and they replicate it along with the plasmid sequences to which they are joined. This is why plasmids used in cloning are called vectors; they are used to ferry pieces of foreign DNA into bacterial cells.

As we just saw, *E. coli* cells can make hundreds of copies of recombinant plasmids containing a human gene. This means that *E. coli* in the laboratory contains hundreds of copies of the human gene you want to express, or, in other words, the dosage of human genes in these bacterial cells is very high. With the appropriate promoter and terminator, these human-gene copies are transcribed at high level (many mRNA molecules are made), and consequently, translation of these mRNAs will generate large amounts of proteins corresponding to that gene. This is possible thanks to the fact that the genetic code is universal. In fact, not only does *E. coli* not "know" that it harbors and actively replicates a piece of human DNA, it is also "unaware" that it is making large amounts of human mRNA and large amounts of the corresponding human protein. Today, large vats, called fermenters, are used to grow engineered *E. coli* on an industrial scale. The protein of interest is purified from these cells and sold by pharmaceutical companies.

Medically Important Human Proteins Made in *E. Coli*

Human proteins manufactured in *E. coli* cells include, among others, insulin, erythropoietin, blood-clotting factors, somatotropin, osteogenic protein, and tPA, or tissue plasminogen activator. Insulin is used to assist individuals who cannot make their own insulin, people suffering from type I diabetes. Erythropoietin is a protein factor that plays a key role in the production of red blood cells. It is used to treat anemia, in particular anemia that may follow cancer chemotherapy. It is also a drug abused by some athletes to increase their amount of red blood cells. Blood-clotting factors are administered to patients suffering from hemophilia, a disease that prevents blood from clotting properly. The osteogenic protein can be used to assist in bone healing. Somatotropin, also called growth hormone, is administered to young people of short stature who do not make enough of this hormone. Finally, tissue plasminogen activator is frequently used in the emergency rooms of hospitals to treat patients who have just suffered a se-

Table 5.1 Examples of Recombinant Products on the Market

Product	Purpose
DNase	Treating cystic fibrosis
Erythropoietin (EPO)	Promotion of red-blood-cell growth, including for patients undergoing chemotherapy
Blood-clotting factors	
Factor VII, VIII, IX	Clotting factors for people with hemophilia
G-CSF	Reduction of infection in cancer patients receiving anticancer drugs
Glucagon	Treating severe hypoglycemia
Glucocerebrosidase	Long term enzyme-replacement therapy for type I Gaucher disease
IL-2 (interleukin-2)	Treating metastatic cancers
Insulin	Managing diabetes
Interferons	Treating chronic hepatitis C and multiple sclerosis
Interleukin	Treating deficiency in blood-clotting cells after chemotherapy
Natriuretic peptide	Assisting heart function in patients with congestive heart failure
Osteogenic protein	Assisting in bone healing
Somatotropin	Promotion of growth for those who normally lack this hormone
tPA	Treating acute myocardial infarctions

Source: http://www.biopharma.com/

vere heart attack. At least a dozen pharmaceutical companies manufacture these and other protein factors, representing a market of several billion dollars (table 5.1).

What is the advantage of making human protein products in *E. coli*? Let us take the example of insulin. Before the advent of recombinant DNA technology, diabetics had to purchase insulin made from pig pancreases collected from the slaughterhouse. Pig insulin is not very antigenic; that is, this protein does not trigger a quick immunological response in humans. However, after years of daily use, a percentage of patients started developing immune reactions toward pig insulin. They then had to switch to sheep insulin, for example. These inconveniences are now a thing of the past; human insulin made in *E. coli* does not trigger an immunological response in diabetics because it is genuine human insulin. In addition, human insulin made in bacteria can never be contaminated with animal viruses because these viruses cannot be propagated in bacteria.

A number of blood-clotting factors provide further examples of human products made by using recombinant DNA technology. The clotting process of your blood involves a cascade of events involving different factors. Each factor is encoded by a separate gene. The hemophilia in Queen Victoria's family is due to a dysfunctional gene located on the X chromosome. There is a second gene on the X chromosome that encodes another clotting factor, and there is a third gene, not on the X chromosome, that encodes yet another clotting factor. These factors are now available for hemophiliac patients via recombinant DNA technology as injectable solutions. Before the advent of recombinant technology, hemophilia patients relied upon plasma from donated blood. Factors from donated blood have the potential of containing dangerous viruses such as HIV or hepatitis. In fact until clotting factors were available via the recombinant DNA technology, 90 percent of adult hemophilia patients were infected with either HIV or hepatitis C.

Finally, other types of cells in addition to bacteria can be used to produce recombinant human proteins of medical importance. Yeast cells, normally used to make bread, beer, or wine, do express human proteins when properly engineered with recombinant human genes. Mammalian cells, including human cells, grown in a culture medium are also used to make recombinant human protein.

The scientists who discovered key tools for recombinant DNA technology, such as restriction enzymes and plasmids, did not know that their work would lead to such important practical applications. This is a beautiful example of the power of science: most of the time, basic discoveries are made without any particular practical goal in mind. However, some basic discoveries can be put to practical use in a very short period of time with important results, in particular for the field of human medicine.

Summary
We have seen that the recombinant DNA technology now makes available significant amounts of many human proteins that were not available before the development of gene cloning. Gene cloning requires the use of restriction enzymes and plasmid vectors. Restriction enzymes cut DNA pieces at specific palindromic sequences leaving short, single-stranded, complementary or "sticky" ends. Plasmid vectors are small circular pieces of DNA found in bacteria that can be used to shuttle DNA pieces into bacteria. Genes to be put into bacteria must have appropriate promoters and terminators so that the bacterial cells that are transformed by the recombinant DNA can make many copies of the gene and the protein that it encodes.

CHAPTER 6

Genetically Modified Plants

YOU MAY HAVE HEARD ABOUT CONTROVERSIES surrounding genetically modified plants. These controversies have even caused street demonstrations and riots in several countries. You may also be assured that genetically modified food plants or their products find their way, on a daily basis, onto your breakfast, lunch, or dinner plates. Thus, it is important to understand what genetically modified plants are and how they are made. This scientific knowledge will allow you to form an informed opinion about the use of genetically modified plants.

What Are Genetically Modified Organisms (GMOs)?

The words "genetically modified" constitute a serious misnomer. Humans have bred plants and animals for thousands of years. In selective breeding, we have built into animals and plants gene combinations that are not normally found in nature and that probably would not survive without human intervention. For example, Chihuahuas and Great Danes did not appear spontaneously, nor can these or other dog breeds be maintained without human intervention. We also know that both breeds are quite different from "primeval" dogs domesticated thousands of years ago. The latter looked like wolves.

Similarly, several types of food plants look quite different from their progenitors. Corn, for example, derives from plants that produced very small purple kernels, not large, yellow ones. Progenitors of modern corn, however, apparently appeared spontaneously from wild parents and were propagated and bred to enhance the characteristics that we enjoy in modern corn. On the other hand, plant breeders mutated the genes of the progenitors of modern barley with chemicals or radiation and selected for mutants with desirable characteristics to produce the high-yielding, short-stemmed strains of barley cultivated today. Thus, we see that humans have modified the genes of the creatures all around us, starting a long time ago in some cases, either by mixing genes from different parents, by selecting interesting variants, or by directly acting on genes through chemical or physical agents. All these plants and animals can legitimately be called "genetically modified organisms" relative to their progenitors.

When the term "genetically modified organism" is used, however, it does not refer to the examples mentioned above. Rather, the term "genetically modified" (GM) is currently applied to plants and animals that result from *adding* genes, in particular genes from completely unrelated organisms, to preexisting plants or animals. In other words, GMOs are regarded as life forms that could not have appeared naturally. This is indeed true for the types of genetically engineered plants that we will discuss in this chapter. Genetic modification of animals, including humans, is discussed in chapter 13.

Genetically modified plants are commonly grown today in the United States (see table 6.1). Most of the corn (including corn products found in breakfast cereals) and soybean- and canola-derived products sold in this country are the result of plants engineered with recombinant genes. The most common genetic modifications are those that confer resistance to certain insect pests and those conferring resistance to certain herbicides. Let us first see how genetic modification of plants is achieved. Two techniques are used to genetically engineer plants: *Agrobacterium*-mediated gene transfer and biolistics.

Agrobacterium-Mediated Gene Transfer
This first method relies on the ability of the soil bacterium *Agrobacterium tumefaciens* to transfer some of its plasmid DNA to plant cells. *Agrobacterium* is a plant pathogen that, after infecting its host, causes the disease called crown gall. This disease is characterized by the for-

mation of tumors at the point of infection. In the infection process, *Agrobacterium* transfers its plasmid DNA to the plant cell, and this DNA is subsequently integrated within the DNA double strand of the infected plant cells. Within this plasmid are tumor-causing genes, so the plasmid is called the Ti (for tumor inducing) plasmid. The important point here is that *Agrobacterium* is a *natural genetic engineer*, since it can naturally transfer its DNA to a plant host. In order to insert foreign genes into plants, scientists remove the tumor-causing genes from the Ti plasmid and insert a foreign gene in their place. The Ti plasmid is thus a natural plasmid vector useful for plant transformation.

A typical plant transformation experiment consists in harvesting plant tissues (for example, leaf pieces) and incubating them with *Agrobacterium* cells containing the gene of interest in its Ti plasmid. After just a few hours, *Agrobacterium* has transferred its Ti plasmid with the foreign genes to the plant cells. Whole plants are then regenerated from these treated plant pieces. This is accomplished by cultivating the plant pieces in a petri dish with a nutrient medium containing plant hormones that promote both root and stem growth. Thus whole fertile plants can be regenerated from the treated tissues. These plants can then be propagated by the usual means, seeds. The complete genetic modification process is shown in figure 6.1.

Biolistics

Biolistics, the other gene transfer technique, does not rely on *Agrobacterium*. Here, a *physical* means is used to force foreign DNA into plant cells. Back in 1987, researchers came up with an idea that must have found its origins in the Wild West. They reasoned that DNA, somehow absorbed on the surface of microscopic metal particles, could actually be *shot* at plant cells by using a gun. In this way, a DNA "bullet" of sorts would penetrate the plant cell wall, and that DNA would become expressed in the plant cell. And it worked! DNA can indeed stick temporarily to microscopic gold or tungsten beads. This shot is loaded into a gun (in early models, the barrel of a .22 caliber was sawed off and the gun welded to a thick metal lid covering a metal firing chamber), and the gun is fired at the plant target. The discharge must be powerful enough for the beads to penetrate the plant cells, but not so powerful as to splatter the sample all over the chamber. You might guess that this was rather tricky to achieve with the explosives used in a regular shotgun. Modern "gene guns," as they are called, accelerate

Leaf with pieces cut out

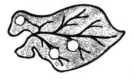

**Agrobacterium shown with
a Ti plasmid with a foreign gene**

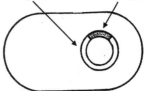

**Pieces of leaf with *Agrobacterium* with foreign gene
in a petri dish**

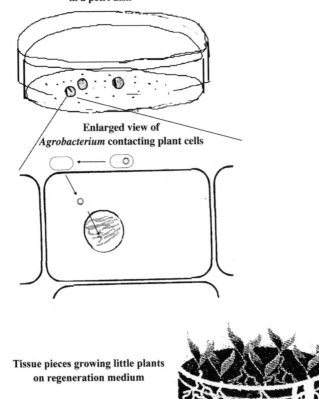

**Enlarged view of
Agrobacterium contacting plant cells**

**Tissue pieces growing little plants
on regeneration medium**

Figure 6.1 Genetically Modifying Plants Using *Agrobacterium*-Mediated Gene Transfer. The process begins by using a piece of plant tissue, for example, a leaf. A gene for the desired trait is cloned into the Ti plasmid of *Agrobacterium*. The pieces of leaf are then placed in a petri dish together with *Agrobacterium* containing the foreign gene. The bacterial cells make physical contact with the plant cells and transfer the plasmid-borne foreign gene to the plant cells' chromosomes. The tissue pieces are then placed on regeneration medium. Each tissue piece will grow into several little plants containing the foreign gene.

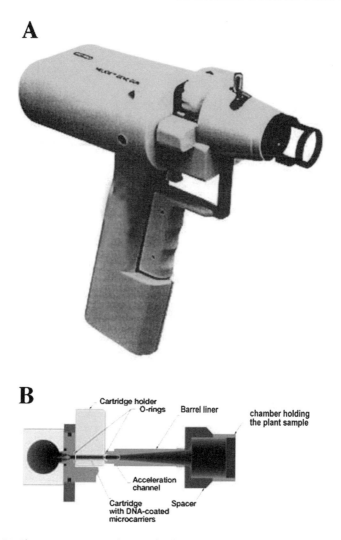

Figure 6.2 The Gene Gun. A. A photograph of one commercially available biolistic instrument called a gene gun. B. A diagram of the gene gun.

metal beads through the quick release of helium gas under pressure (figure 6.2). Basically, once the DNA-coated beads find themselves inside the plant cells, DNA is released from them, migrates to the cell nucleus, and ends up integrated within the plant DNA. Biolistics thus achieves the same result as *Agrobacterium*-mediated gene transfer,

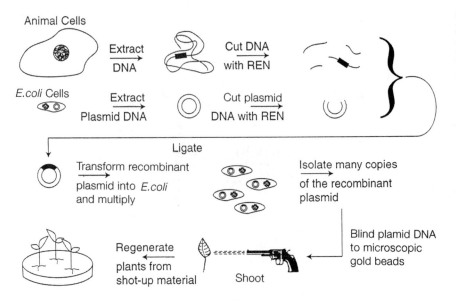

Figure 6.3 The Biolistic Process of Genetically Modifying Plants. First, DNA is isolated from the cells of an organism. Plasmid DNA is isolated from *E. coli*. Both the plasmid and the DNA from the other organism are cut with the same restriction enzyme (labeled REN). The cut pieces of DNA and plasmid are mixed and ligated (glued) back together. The plasmids are introduced into *E. coli* to make many copies of the recombinant plasmid. The recombinant plasmid containing the desired gene is isolated and attached to gold beads. The beads are loaded into a gene gun and shot into plant tissue. The tissue is then placed on growth medium as in the *Agrobacterium* method.

without the use of live bacteria, however. As in the case of *Agrobacterium*, shot-up plant tissues must be regenerated into fertile plants. The biolistic process is summarized in figure 6.3.

Note that when the foreign genes used in these experiments are of a nonplant origin, they must be equipped with a promoter and a terminator that function in plants. A commonly used promoter is derived from the cauliflower mosaic virus (CaMV), a virus that infects cauliflower, broccoli, and other members of the cabbage family. A commonly used terminator is called a "nos terminator," a terminator for a plant-hormone gene found in *Agrobacterium* Ti plasmid. These specific pieces of DNA are cut and put together using the recombinant DNA techniques explained in chapter 5.

Box 6.1 *Detecting Foreign Genes in Genetically Modified Plants*

For a variety of reasons, it is important to know whether plant products are genetically modified. For example, some farmers grow and specifically sell non-GM crop plants to their customers; some countries require that only non-GM products be imported; other countries require labeling products containing GM plants. Since GM and non-GM plants or products made from them cannot be distinguished by outward appearance, genetic tests are necessary to determine whether a product is or is not genetically modified. Since it is not always known what foreign gene might have been used to engineer a particular plant, a genetic test was developed based on the detection of the promoter of the foreign gene.

Foreign genes, if they are not of plant origin, must be engineered with plant-expressible promoters for proper functioning in a plant host. Cauliflower mosaic virus promoter, isolated from a virus that infects plants in the cabbage family, is commonly used because it is very active in many different plants. Thus, regardless of the nature of the foreign gene itself, it is possible to distinguish between GM and non-GM plants if the genetic engineer used the CaMV promoter. Detection of the CaMV promoter is done by polymerase chain reaction (PCR; see chapter 1). Since the sequence of the CaMV promoter is known, we can design primers for it. We can test the DNA isolated from plants or plant products using these primers in a PCR reaction. The length of the CaMV promoter is 195 base pairs. If the promoter is present in a plant product, the PCR reaction will amplify a 195-base-pair piece of DNA that can be easily detected by gel electrophoresis (chapter 1). Of course, if the CaMV promoter was not used, this technique will not determine whether or not the plant was genetically modified. However, if the promoter is detected, and the cauliflower mosaic virus does not normally infect the tested plant, one can be quite confident that there was some genetically modified plant product in the sample.

Genetic Modifications

The most common GM food plants on the U.S. market are insect-resistant corn, herbicide-resistant soybeans and canola, and virus-resistant papaya. Table 6.1 lists all the genetically modified plants that have been approved as of 2002 by the U.S. Food and Drug Administration.

Let's first consider genetic modification for insect resistance. The gene for insect resistance is isolated from the bacterium *Bacillus thuringiensis*, a natural pathogen of certain insects. This gene codes for a toxin, called Bt toxin, that kills the insects' intestinal cells. Insects infesting plants with this Bt toxin gene ingest the toxin as they feed, and they die. Corn, potato, and cotton plants have been engineered with this gene. Interestingly the Bt toxin gene in use today is synthetic, just like the insulin gene discussed in chapter 5. That is, the amino acid sequence of the toxin protein was decoded into the corresponding DNA sequence, and that sequence made in the test tube.

Two examples of herbicides for which plants have been genetically modified for resistance are Roundup® (produced by Monsanto) and Liberty® (produced by AgrEvo). Both herbicides are kill-all, meaning that they kill all plants. Herbicide resistance can be achieved by inserting genes for enzymes that have the ability to break down the herbicides. Such genes have been isolated from bacteria. Thus, plants engineered for resistance to either herbicide will survive being sprayed by the herbicide, while any plants without this gene, including weeds, will not.

Interestingly, resistance to viral infection has been achieved by inserting into plants one of the genes of the virus itself. Viruses are typically composed of a protein coat and a DNA or an RNA genome. A generic virus can be visualized as a molecule of nucleic acid "wrapped" into a layer of proteins and sometimes lipids. The shape of the protein coat is extremely variable and is virus-dependent. Think of this protein coat as "packaging" the viral genome and containing the necessary ingredients to deliver it inside host cells. For reasons not entirely clear, if a plant is genetically modified with the gene for the coat protein of the virus, the genetically modified plant exhibits resistance to infection by that virus.

All the examples discussed so far deal with plants engineered for resistance to some kind of injury, inflicted either by viruses, insects, or herbicides. Some researchers, however, have attempted to and succeed-

Table 6.1 Genetically Modified Crops Approved by the Food and Drug Administration Through 2002

Purpose	Source of Gene	Crops
Delay ripening	tomato	tomato
	bacteria	tomato
	bacteria	cantaloupe
	bacteriophage	tomato
Resistance to herbicide	bacteria	soybean
	bacteria	cotton
	bacteria	canola
	bacteria	corn
	bacteria	radicchio
	bacteria	sugar beet
	bacteria	rice
	plant	flax
	plant	corn
Resistance to insect pests	bacteria	potato
	bacteria	cotton
	bacteria	corn
Resistance to viral pests	virus	potato
	virus	papaya
	virus	squash
Male sterility and restoring fertility	bacteria	corn
	bacteria	canola
	bacteria	radicchio
Nutritional change	plant	canola
Feed-quality change	fungus	canola

Note: although the above have been approved, they are not necessary in commercial production.

ed in changing the nutritional quality of canola by changing its oil composition. Another early example of changing the quality of the food is delayed ripening in tomatoes, as in the case of a tomato called the Flavr-Savr Tomato. Delayed ripening can be achieved by different genetic modifications. One approach is to reduce the activity of the enzyme that produces ethylene, a natural ripener. Another method is to reduce the activity of an enzyme that softens fruit by breaking down pectin. Delayed-ripening tomatoes did reach the grocery stores in the late 1980s. However, they were a market flop and are no longer available.

A genetic modification in progress aims to improve the nutritional qualities of rice. Polished, or white, rice, which is commonly consumed in Asia and elsewhere, contains no vitamin A. Vitamin A deficiency is a serious problem for people whose diet is based mostly on rice, such as poor people in Asia. It can lead to blindness in children and exacerbates other pathological conditions. A Swiss/German university team decided to engineer rice plants to make them able to produce vitamin A (actually provitamin A, which is converted to vitamin A in the human body) in their seeds, the edible part of rice. Rice is actually able to synthesize compounds that can be metabolized into provitamin A. The problem is that rice does not contain the proper enzymes to do this in the seeds. The researchers isolated two genes coding for the relevant enzymes, one gene from the soil bacterium *Erwinia ureidovora* and another gene from daffodils. Both genes were transferred to rice plants using the *Agrobacterium* method, and, sure enough, the rice grains started making provitamin A, giving the grains a golden color, the natural color of provitamin A. For this reason, this rice variety has been dubbed "golden rice." Golden rice is still in its developmental phase. In order to produce commercial rice that makes provitamin A, the transformed plants must be regenerated from tissue cultures and then crossed with varieties that have the growth and edibility qualities favored by farmers and consumers. Field trials of these golden-rice hybrids are underway.

GM crop plants are not grown in Europe or Japan. In both Europe and Japan, there is strong opposition to GM plant products. The European Union has had a ban on foods containing GMOs. They are now considering new rules to allow foods containing GMOs if they are labeled as such. However, in China sixteen GM plant varieties are either in production or soon to be. These varieties include rice, wheat, potato, corn, soybean, canola, peanut, cabbage, tomato, chili, and others, genetically modified for insect, virus, or herbicide resistance. Given all the controversy surrounding GM food plants, it is unclear what their future will be.

Genetically Modified Nonfood Plants

Not all potential applications of genetic modification of plants concern food plants, however. In this section, we will describe two other potential applications of genetically modified plants: phytoremediation and human vaccine production in plants.

Phytoremediation is the act of cleaning up polluted soils with plants. As we know, vast stretches of land, worldwide, are heavily polluted with petrochemicals or toxic heavy-metal salts such as mercury and cadmium. It so happens that soil microbes are often able to detoxify these pollutants by degrading petrochemicals and immobilizing heavy metals. However, there are potential problems with spraying contaminated soils with these microbes, for fear of creating yet another environmental problem. This is where GM plants may come to the rescue. Why not isolate the relevant bacterial genes and transfer them to plants? Plants do not move, can grow fast, and, through their roots, absorb polluted water, thereby concentrating toxic chemicals inside their cells. When engineered with bacterial genes, these GM plants would be able to survive. After their detoxification job was achieved, they could be harvested and burned under controlled conditions. And yes, this has been done. There now exist fast-growing GM poplar trees able to detoxify petrochemicals and salts of various heavy metals. Their positive effects have been tested on a small scale, and research is being pursued for other pollutants using poplar trees as well as other plants.

Finally, plants, one day, may be a source of readily accessible human vaccine. Vaccines are made from disabled pathogens, such as viruses or bacteria, so that they can no longer cause disease but instead provoke an immune reaction and thus provide protection from the pathogen. When people are injected with a part of a pathogen called an antigen, their immune system produces specific antibodies that block the antigen. Vaccinated people, when exposed to the pathogen, will also block it, thanks to their newly acquired antibodies. Vaccination works very well to avert a host of viral and microbial diseases. Vaccines, though, typically require refrigeration and must be injected using sterile syringes. These are not always easy to come by in developing countries: hence the idea of producing vaccines in edible plants. This concept has been successfully demonstrated: a gene from the Newcastle virus (a pathogen that causes severe diarrhea) has been cloned and expressed in potatoes. Volunteers who ate these GM potatoes developed antibodies against the virus. These GM potatoes had to be eaten raw, because boiling or frying would have destroyed the antigen. Since most people would not enjoy chewing on raw potatoes, the banana is now being considered for genetic modification with antigen genes. Vaccination campaigns with GM plants have not yet started, however. Amusingly, since vaccines are pharmaceuticals, the idea of producing them in plants has been dubbed "pharming." Clearly, other proteins of medical importance could potentially be produced in GM plants. One great advantage is that plants cannot be contaminated with human viruses, while plant viruses are harmless to humans.

Ecological Issues

Well before the introduction of GM plants, some people voiced concerns about the impact these plants could have on the environment. We will consider two major ecological issues associated with genetically modified plants: the appearance of resistance to Bt toxin among target insects and the possible creation of "superweeds."

When insect populations are sprayed with an insecticide, resistance to this chemical invariably occurs. One famous example is DDT. This insecticide has completely lost its effectiveness in many parts of the world because so many insect species (including mosquitoes that carry the malaria parasite and the tsetse fly that carries the parasite causing sleeping sickness) have developed resistance to it. Fears that insects that are pests for cotton would develop resistance to the Bt toxin made by GM cotton have been borne out. Just two years after the introduction of cotton modified with the Bt toxin gene, such resistance has appeared in the field.

We must minimize the rise of resistance if Bt toxin is to remain an effective tool to fight pest damage. One such measure, instituted since the initial introduction of Bt modified crops, is that farmers who plant Bt crops are now required to grow a certain percentage of non-GM crops in close proximity of the GM crop. The idea is to create refuges in which susceptible insects can survive, that is, areas planted with crops that do not contain Bt toxin. This should help to maintain insect populations still susceptible to the toxin. Indeed, if Bt crops were the only food source for insects, only resistant individuals would survive and reproduce. Because susceptible individuals are not killed by Bt toxin in the refuges, availability of the refuges reduces the rise of resistant individuals. The goal of having refuges is to allow many sensitive insects to survive and dilute the effect of rising resistance. We cover more about the rise of resistance in a population in chapter 11. The lesson with Bt-modified crops is that genetic modification of crops with Bt toxin is not a cure-all against insect infestation of crops.

Another potential ecological problem is the transfer of genetically modified traits from GM crops to their weedy relatives. For example, if herbicide resistance in crop plants is transferred to their weedy relatives, it can produce "superweeds." All cultivated plants are derived from plants found in the wild. Many crops still have wild, weedy relatives with which they can occasionally interbreed. If this breeding happened between herbicide-resistant GM crop plants and their weedy relatives, herbicide-resistant weeds would be created, and they could destroy the effectiveness of the herbicides. Another example is that of transfer of viral resistance from genetically modified crops to their weedy relatives. For example, viral resistance in squash may be transferred to their wild cousin that is a pest, especially in the Southern US. The transfer of genetic modification from GM crops to their wild relatives has not yet been documented in the wild. Yet it is clear that a small amount of interbreeding can occur between domesticated crops and their wild counterparts.

Labeling Issues and Food Safety

Let us now discuss labeling issues and food safety. People are legitimately concerned about the safety of what they eat. Even though Bt toxin is deadly for insects, it has no effect on human intestinal or other cells, and it is not allergenic because it is quickly destroyed in human gastric juice. Similarly, enzymes that degrade herbicides have been tested for toxicity and allergenicity. GM food plants have been extensively analyzed for the presence of toxic compounds and have been approved by the Food and Drug Administration. One test for transgenic products is to determine whether the foreign protein is degraded in the stomach. If the protein is not degraded in the stomach, it has the potential for triggering allergic reaction. Nevertheless, not everybody is convinced that GM food plants are entirely safe.

One solution to this question of food safety is to label foods that contain GM plant products. Indeed, the European Union is considering allowing the importation of products containing GM plants if the products are labeled. If this were done, the public could choose between GM and non-GM foods. There are two obstacles to labeling genetically modified foods: One is that biotech companies have fought hard to prevent the labeling of these products in the United States, and so far they have won this battle. This means that the public has no way of knowing which foods contain GM plant products and which do not. Another looming obstacle is the fact that so many staple crops in the United States are genetically modified. If processed foods are to be labeled, even if only a small portion of the contents is genetically modified, a major transformation of farming and distribution practices in the United States is necessary to keep genetically modified crops separate from nonmodified crops in the food stream.

This issue of labeling of genetically modified foods promises to be contentious for years to come. The increasingly global market of foods will further heighten the need to be aware of issues related to the genetic modification of crops.

Summary

We have seen how plants can be genetically modified by the addition of genes from another organism. Two methods are commonly used to introduce foreign genes into plants. The first method uses a natural genetic engineer in plants, the bacterium *Agrobacterium tumefaciens*. For this, the foreign gene is first inserted into *Agrobacterium*'s Ti plasmid. Then the *Agrobacterium* naturally transfers the recombinant plasmid into the plant cells when the two are mixed. Another technique, called biolistics, consists of bombarding plants with microscopic, DNA-coated metal particles. Currently, genetically modified plants are commercially produced and marketed in the United States. Ecological, food safety, and labeling concerns surround the growing and selling of genetically modified plants.

CHAPTER 7

When Things Go Wrong

SO FAR, WE HAVE LEARNED THAT GENES RESIDE on chromosomes as sequences of DNA base pairs. When an organism develops from a single fertilized egg, each time a cell divides, the DNA is faithfully copied (chapter 1) and the billions of base pairs that are organized into chromosomes are divided equally between daughter cells in a process called mitosis (chapter 2). When we pass our genes to our offspring, DNA in specialized cells (gametes) is faithfully copied again, but these cells divide twice in a process called meiosis (chapter 2). This process is necessary in order to halve the genetic material in the gametes. Thus, the gametes contain only one copy of each chromosome. In this manner, we create a new individual by joining one copy of the chromosomes from the mother and one copy of the chromosomes from the father, and each offspring ends up with two copies of each chromosome in the fertilized egg. This is what usually happens. Occasionally, however, things go wrong when chromosomes separate during meiosis or when DNA is copied. Errors in these processes are called mutations. They can have little effect or they can have major consequences on the phenotype of a person.

Errors in Chromosome Number

We are all products of a sperm fertilizing an egg. Humans have twenty-two pairs of chromosomes, plus the sex chromosomes. After meiosis, the sperm should have twenty-two chromosomes plus either the X or Y chromosome, and the egg should have twenty-two chromosomes plus an X chromosome (see figure 7.1). But sometimes, errors occur in the process of meiosis, and a given gamete is either missing a chromosome or has more than one copy of a particular chromosome. . When this happens, and the egg or sperm that has an aberrant number of chromosomes is involved in making a new individual, of course, that individual will have a wrong number of chromosomes. Because there are so many genes on each chromosome, such errors would be expected to be very deadly. A great majority of fetuses with abnormal chromosome numbers do not survive to birth. Thus, chromosomal abnormalities explain some cases of infertility and pregnancy loss. Indeed, up to 50 percent of first-trimester spontaneous abortions are due to chromosomal abnormalities. A small percentage of fetuses with chromosome abnormalities do survive to birth, and they account for a large proportion of babies with congenital malformations and mental retardation. The most common abnormality involving an abnormal chromosome number is Down's syndrome. This condition is due to an extra copy (for a total of three) of chromosome twenty-one, so it is also called trisomy 21. Individuals with Down's syndrome are typically short and have lower IQs than average, but they can live productive lives. Other relatively common cases of chromosomal abnormalities involve the sex chromosomes. Because the Y chromosome is not essential, and we only need one X chromosome, variations in sex chromosome numbers can occur without lethal consequences. Individuals with unusual sex chromosome numbers (such as XXX, XXY, XYY, and XO, where the individual has just one X chromosome) may be sterile or may show anomalies in secondary sexual characteristics that are generally not noticeable.

Multiple Sets of Chromosomes

Chromosomal abnormalities can also occur if more than one sperm fertilizes an egg or if the developing embryo fails to divide its cells properly. These situations result in the addition of a complete set of chromosomes. Normal individuals with two sets of chromosomes are called diploid, while those with three sets are called triploid, and those with

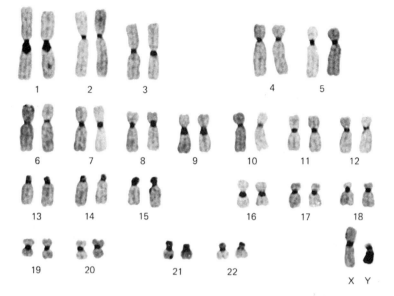

Figure 7.1 Human Male Karyotype.

four are called tetraploid. These conditions in humans and other mammals are so deleterious that affected fetuses do not survive to birth.

Interestingly, however, extra sets of chromosomes are not deleterious to many other organisms, including all plants and some animals, including fish. Indeed, many crop plants we eat are not diploid. In order to reproduce, the reproductive cells must undergo meiosis, a process in which they double their chromosome numbers and then divide them equally so that the gametes contain half the number of chromosomes (see figure 2.3). In the case of diploid individuals, meiosis results in the presence of a single set of chromosomes, a haploid chromosome set, in all the gametes. Plants that have more than two sets of chromosomes, if the number of sets is even, can still reproduce because they can evenly divide their chromosomes. For example, tetraploid plants produce diploid gametes, and octaploid plants produce tetraploid gametes. Strawberries that we buy at the grocery store are octaploid. Similarly, wheat is hexaploid: it contains six sets of chromosomes, an even number of chromosome sets that can be evenly divided into triploid gametes.

What, however, is the effect of triploidy, the presence of three sets of chromosomes in an organism? An example of a triploid plant is the banana. All commercial bananas are triploid. In this case, there is no way to divide the three sets of chromosomes equally in halves. Remember that during meiosis, the chromosome sets are first doubled. In the banana, the number of sets would then be six. But then these six sets of chromosomes would have to be divided in half twice, or into four, since each pregametic cell gives rise to four gametes. The result is one and one-half sets of chromosomes, not a whole number, and thus meiosis cannot take place properly. Triploid commercial banana trees are sterile; their fruits do not make seeds. On the other hand, wild bananas, which are diploid, do have seeds (see figure 7.2). Thus, commercial bananas are reproduced by cuttings. Using similar logic, the fish and game departments in many states purposely make triploid fish, so that the fish they release for the fishing public are unable to breed and reproduce with wild fish. An additional bonus in triploid fish is that the energy a diploid fish invests into reproduction is used for growth in triploids, resulting in bigger fish.

Looking at Our Chromosomes

We can detect chromosomal abnormalities by organizing chromosomes into a karyotype. A karyotype is a display, arranged in an orderly fashion, of all the condensed chromosomes present in a dividing cell. Figure 7.1 shows the karyotype of a man. If we could look into our cells with a microscope, most of the time our chromosomes would not look like those shown in the figure. For one thing, the chromosomes are so tightly packed that it is difficult to distinguish them. For another, the chromosomes condense and become visible under the microscope only when the cells are about to divide. In order to observe our karyotype, we need to induce cells to divide. To obtain a karyotype of an adult, white blood cells from a blood sample are used. In the case of a fetus, cells from the amniotic fluid surrounding the fetus or chorionic villus cells can be used (figure 7.3).

The sampled cells are transferred into a culture flask, a chemical that promotes cell division is added, and the cells are allowed to begin dividing. Then a second chemical is added that stops the cells just as the chromosomes are fully condensed. The cells are then mixed with a fluid that makes them swell and renders them very fragile. The cell suspension is subsequently dropped onto a microscope slide, with the result that the bloated cells burst open. The condensed chromosomes are released and spread out on the glass slide. After staining, they are

Figure 7.2 Banana Seeds. Picture of wild banana seeds next to a commercial banana, which do not have seeds because they are triploid.

photographed under the microscope. Using image-analysis software, the chromosomes are ordered, from the largest, chromosome 1, to the smallest, chromosome 22. The sex chromosomes are usually grouped separately (figure 7.1). Extra chromosomes are easily detected, as in figure 7.4, which shows the karyotype of an XXY human. Individuals with XXY karyotype are males with reduced fertility and with some secondary female sex characteristics. Down's syndrome is diagnosed in the same way, with the detection of an extra copy of chromosome 21. We will see next that genetic abnormalities can be the result of much more subtle changes in the DNA, and not the consequence of adding or losing whole chromosomes.

Changes in the DNA Base Sequence

Our genetic information is contained in the precise ordering of the DNA base pairs. Although the DNA-copying process is very accurate, rare errors do occur. As in computers that are equipped with an error-checking system, cells possess mechanisms that minimize mistakes in DNA replication. Yet each time our cells divide, all 3 billion base

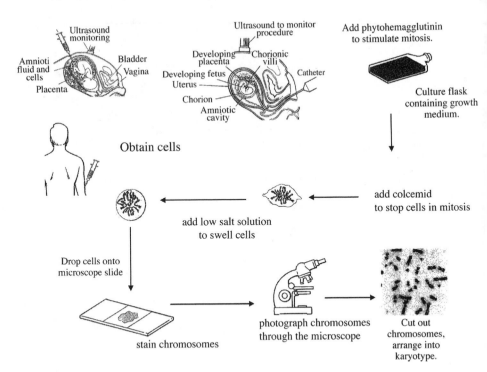

Figure 7.3 How Karyotypes Are Made. First cells are obtained from blood or other cells of an adult or a fetus through chorionic villus sampling or amniocentesis. If necessary, for example, with adult cells, phytohemagglutinin, a chemical that stimulates growth and division is added. Then, after growth, colcemid, a chemical that stops the cells in mitosis is added. Then these cells are swollen in a low-salt solution and dropped onto a microscope slide. This step bursts open the cells and spreads out the chromosomes. The chromosomes are then stained and photographed through the microscope. In the past, photographs of chromosomes were literally cut up with scissors and arranged by size and shape. Now, this is all done with image-analysis software.

pairs must be faithfully copied. Given this huge number, even with an extremely accurate DNA copying process errors are still bound to occur. What type of errors can occur and what are their effects?

One type of error is the substitution of one base for another during copying. There are many different possible outcomes from such an error. For example, if the gene sequence is

. . . . . . CTT TGC AGT GCC CTC CAG AAA ATA AAG TAA

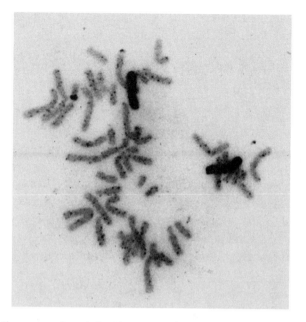

Figure 7.4 Chromosome Spread of an XXY Human. This chromosome spread was made using a special technique that highlights the sex chromosomes, making them darker than the other twenty-two pairs of chromosomes. The Y chromosome is the dark spot towards the top left, the two dark larger chromosomes are the X chromosomes. Photo courtesy of Norah McCabe.

it codes for a protein with the following amino acid sequence (see figure 4.6):

......L C S A L Q K I K

If a DNA replication error results in G (in bold) at the third nucleotide of the first codon rather than T, which is the correct nucleotide,

......CT**G** TGC AGT GCC CTC CAG AAA ATA AAG TAA

there will be no difference in the resulting amino acid because both CTT and CTG code for the same amino acid, leucine (see the genetic code in figure 4.6). This change is called a base substitution, but here there is no amino acid substitution.

On the other hand, if an error results in the following substitution at the first nucleotide position,

......ATT TGC AGT GCC CTC CAG AAA ATA AAG TAA

then isoleucine, instead of leucine, will be inserted in the growing protein. Therefore, this base substitution results in an amino acid substitution. But, because leucine and isoleucine are very similar chemically, in many cases there will be no detectable difference in the protein function and thus no change in phenotype.

But, if the error results in a substitution at the second nucleotide position of the same codon,

...... CGT TGC AGT GCC CTC CAG AAA ATA AAG TAA

then arginine will be substituted for the leucine. These two amino acids are very different chemically. So, depending on the part of the protein that is affected, this error may result in a major change in function or make the protein nonfunctional.

Finally, there may be an error such as the following:

...... CTT TGC AGT GCC CTC TAG AAA ATA AAG TAA

Since TAG is one of three stop codons, the protein will end there rather than at the original stop codon TAA. Often such an error results in a shortened, nonfunctional protein. However, if the error occurs near the normal end of a large protein, as in this case, it may still result in the making of a partially functional protein. Therefore, the effect of changes in the base pairs of a gene can range from no effect at all to causing a major alteration that results in a totally nonfunctional protein.

The base sequence of DNA can be modified in other ways. Occasionally, an additional base may be inserted within a normal sequence. Alternatively, one of the bases may be skipped. These are called insertions and deletions, respectively. Let us look at the effect of the insertion of a base in the second codon of the following sequence

...... CTT TGC AGT GCC CTC CAG AAA ATA AAG TAA

which codes for the following sequence of amino acids.

...... L C S A L Q K I K

Let us now insert an extra T (in bold) after the third T in our original sequence. We then get

...... CTT **TT**G CAG TGC CCT CCA GAA AAT AAA GTA A

this sequence now codes for a protein with the following amino acid sequence.

...... L L Q C P P E N K V

Note that from the point of insertion on, all the amino acids are different from the original ones! But more devastating than this, though this is a major problem in and of itself, now there is no stop codon in this sequence anymore! Thus this single base insertion results in a much longer protein than normal, with totally different amino acids! Similar situations occur in the case of a deletion. Try deleting one of the bases near the beginning of the first sequence above. Then use the genetic code in figure 4.6 to determine the amino acids of the mutated protein. Does your deletion mutation also make a larger protein by eliminating the stop codon?

An example of a DNA base change that results in a major defect is Marfan syndrome, a dominant trait discussed in chapter 3. Numerous base additions or deletions in a specific gene result in Marfan syndrome. The gene that is defective in Marfan syndrome, fibrillin, is very large, and most of that gene is important for its proper function. Because of its large size, mistakes can occur over a long sequence of DNA; that is, the gene presents a large target for mutations. Thus many defects in Marfan syndrome are new mutations rather than ones inherited by affected individuals. One example is a CGC to CCC change that results in a change from the amino acid arginine to the amino acid proline. There are other examples of single amino acid substitutions in the large protein that causes Marfan syndrome.

Amino acid substitutions can also affect the function of hemoglobin, the protein responsible for carrying oxygen in our blood. As we have already seen, sickle-cell anemia is caused by glutamic acid, present in normal hemoglobin at the sixth position, being changed to valine. However, the amino acid right next to that normal glutamic acid is again glutamic acid in the seventh position. Variants have been found at that position, where either AAG, which codes for lysine, or GGG, which codes for glycine, replaces the normal GAG-coded glutamic acid. Interestingly enough, these hemoglobins function normally, unlike sickle-cell hemoglobin. Farther down the hemoglobin protein, in position 145, the normal amino acid is tyrosine. In two different mutations, either histidine or aspartic acid replaces this amino acid. Tyrosine, histidine, and aspartic acid are very different chemically, but instead of causing a defective protein, one might say that these mutations improve it, because these mutated hemoglobins have higher affinity for oxygen. These mutations have been found in individuals living at higher altitude, where higher affinity for oxygen improves survival. Table 7.1 shows some other single amino acid changes in hemoglobin and their effects.

Thus, a change in DNA sequence can result in a change of amino acid, but the effect of this change can be dramatically different depending upon the position of the amino acid and the amino acid that replaces it. Some changes are totally unnoticeable, while others cause devastating diseases. Others yet actually "improve" the function of the protein.

Triplet Repeat Errors

A final type of error is referred to as triplet repeat expansion. The genes susceptible to these errors have repeats of triplet nucleotides, that is, sequences of three given bases repeated several times in a row. An error in copying this type of gene makes the triplet repeats longer than they are in normal individuals. A well-studied case of this type of mutation is myotonic (or muscular) dystrophy. A normal individual has from five to twenty-seven copies of the CAG triplet repeat. Individuals with mild symptoms of the disease have at least fifty copies. Severely affected individuals can have over 2,000 copies of this triplet!

Other types of CAG repeat diseases also occur, but can be called GCA or AGC repeats since these are permutations of CAGCAGCAG. Other examples of triplet repeat expansion errors are Huntington's disease and Fragile X syndrome. Normal individuals have nine to thirty-seven repeating CAG triplets in a gene, but it is repeated upwards of one hundred times in Huntington's patients.

Although the cause of triplet repeat error is not known, it is helpful to think of it as a "stutter" during the copying of the DNA. When there are multiple copies of a repeat, the copying machinery can get stuck, or stutter, and put in more copies than there were originally. Once this process begins, it tends to cause even more additions of the triplet repeat in subsequent generations. The tendency for triplet repeats to lengthen, as a result of more copying errors, may explain the fact that offspring of mildly affected individuals are more severely affected, with greater numbers of triplet repeats. These offspring also tend to show the disease phenotype at a younger age. The phenome-

Table 7.1 Examples of Single Amino Acid Variants in β-hemoglobin and Their Phenotypes

amino acid position	6	7	58	66	94	102	144	145
normal	glu	glu lys gly	tyr	glu	lys	asn	lys	tyr
sickle cell	val							
malaria resistance	lys							
high O₂ affinity							asn	asp his
low O₂ affinity			his			ser		

non of increased severity and earlier onset of the disease as the disease is passed down, along with dominance, are characteristics of all triplet repeat expansion diseases studied to date.

Summary

Every time a cell divides, it must accurately copy all the DNA it contains, then divide the duplicated chromosomes equally into daughter cells in mitosis or into four gametic cells in meiosis. The process of copying DNA then dividing the duplicated chromosomes, while very accurate, is not perfect. Errors in the number of chromosomes can lead to major problems in the fetus. Errors that result in multiple sets of chromosomes are also fatal in mammals. However, multiple sets of chromosomes are not lethal in plants and in other animals, such as fish. In fact, multiple sets of chromosomes are common among plants and have been used by plant breeders to produce larger fruits. Errors in copying DNA can result in nucleotide substitutions. These substitutions can have results ranging from no change to very slight changes in the amino acids of a protein to major deleterious changes, or even to an advantageous change. Errors in copying DNA that result in insertion or deletion of nucleotides result in major changes to amino acid sequences because insertions and deletions totally change the amino acids from that point on. Finally, some diseases are due to triplet repeat expansion.

CHAPTER 8

Mutagens, Teratogens, and Human Reproduction

IN CHAPTER 7 WE LEARNED that things can go wrong in the process of passing our genetic information to our offspring. So why do things go wrong? Is there something we can do to prevent mutations? In this chapter we will learn about different reasons for error in passing on our genetic information. We will also learn that nongenetic causes can cause problems in our offspring.

Spontaneous Mutations

DNA polymerase replicates DNA by making complementary copies of each of the double strands. If done perfectly, this process produces exact copies of the original double-stranded DNA. However, nothing is perfect. Consider that we have over 3 billion base pairs of DNA in each cell. Each time a cell divides, whether to produce more of the same cells, like our skin or blood cells by the process of mitosis, or to make sperm and egg through meiosis, all 3 billion base pairs must be replicated. Given this enormous number, DNA polymerase occasionally makes mistakes. Because it is so important to make as few mistakes as possible in copying our genetic material, cells possess proofreading enzymes that "check" newly replicated DNA. These proofreading enzymes make sure that the newly made double-stranded DNA molecules have complementary base pairs: A facing T and G facing C. But in spite of the fidelity of the DNA polymerase enzyme and the action of the proofreading enzymes, rare errors remain in newly replicated DNA. It is estimated that during the replication process, there is approximately one wrong base in one million to one hundred million bases incorporated into DNA. That is a very low percentage of error, but the process is still not perfect. So, just in the normal process of copying DNA, errors, or mutations, accidentally occur. *This is an important concept: mutations can take place without anything being out of the ordinary.* This phenomenon is called spontaneous mutation. Although the spontaneous mutation rate is low, it is not negligible.

Another source of mutation that is often categorized as spontaneous mutation is oxidative damage. You may have heard that vitamins C and E and other foods are antioxidants and that people take them to reduce oxidative damage to their cells. As its name implies, oxidation involves oxygen. Our cells require oxygen to live, yet oxygen is a very reactive molecule and can be damaging to chemicals in our body. For example, oxidative damage changes guanine in our DNA to a chemical called oxoguanine. Because oxoguanine can pair with A, adenine, oxidative damage changes a G-C base pair in DNA to an oxoG-A pair, which after replication ends up as a T-A base pair. Proofreading enzymes can, however, spot and repair oxoG.

The two types of spontaneous mutations described above occur just as a fact of life. That is, there are no extraneous factors that cause these mutations: DNA polymerase makes errors, we cannot totally avoid oxidative damage, and not all errors are caught by proofreading enzymes.

Mutagens

Since mutations occur in the natural process of DNA replication, how shall we define mutagens? It would be strange to define the natural replication of DNA as mutagenic. Thus we will define mutagens as *factors that increase the rate of mutation over and above that of the spontaneous mutation rate.* Two categories of mutagens will be discussed. One is electromagnetic radiation and the other comprises various chemicals.

First, let us look at electromagnetic radiation. Electromagnetic radiation ranges from radio waves to X rays and gamma rays. Our eyes see just a small portion of this spectrum, the visible wavelengths of light.

Microwaves are used in microwave ovens, while infrared radiation can be thought of as heat waves. These two types of electromagnetic radiation are not mutagenic. Electromagnetic waves shorter than visible light, including UV, or ultraviolet rays, gamma rays, and X rays are mutagenic: they increase the mutation rate above the spontaneous rate.

For example, we know that UV light causes specific damage to DNA in the form of thymine dimers. These occur when two Ts adjacent to each other in a DNA molecule fuse together under the effect of UV light. Thymine dimers deform the DNA double helix and greatly increase the possibility of errors in DNA copying. Our cells possess two different repair mechanisms to minimize errors due to thymine dimers. Individuals with defects in either of these repair mechanisms are highly sensitive to UV radiation and are prone to skin cancer. Understandably, there is much concern about increased exposure to UV rays. For example, recent studies show that people using tanning salons increase their chances of getting skin cancer. The shrinking ozone layer is a related concern because ozone helps to prevent harmful UV rays from reaching the surface of the earth. As this ozone layer becomes depleted, we may expect mutation rates to increase in many organisms.

X rays, such as those used in medical diagnosis, are also mutagenic. Patients exposed to X rays use protective lead shields on parts of the body that are not being X-rayed to minimize exposure to the mutagenic effect of X rays. For example, women getting mammograms must be careful to shield their reproductive organs.

Radioactive materials, natural or manmade, are mutagenic as well. All living things that were exposed to radioactive fallout from atomic bomb testing, the atomic bombs dropped on Hiroshima and Nagasaki in 1945, or from nuclear power plant accidents such as Chernobyl (in Ukraine) in 1986, experienced increased mutations if they survived. It is estimated that over 3.5 million people were contaminated with radioactive fallout from the Chernobyl accident. Many different radioactive compounds were released in these and other accidents, but of greatest concern to people is radioactive iodine, ^{131}I. Radioactive iodine is particularly dangerous because the human body accumulates it in the thyroid gland. In fact, a sharp increase in thyroid cancer among children in Ukraine and Belarus was detected four years after the Chernobyl accident, with some areas having thirty to one hundred times the expected rate of thyroid cancers. There are companies selling iodine pills in the form of potassium iodide (KI) that one can take to block the uptake of radioactive iodine that might be accidentally released. Some public agencies are urged to keep a stock of iodine pills in case of emergency.

Chemicals can also be mutagenic, either because they directly damage DNA or because they may be similar in structure to a nucleotide and so can fool the cellular machinery responsible for DNA replication. A chemical in the first category is nitrous acid. We do not eat nitrous acid, but chemicals that are often used in food processing, nitrates and nitrites, are converted into nitrous acid by the acid in our stomach. Once nitrous acid is made in this way, it can convert a C to a U. U is normally only found in RNA, but when it is formed by nitrous acid in the DNA, a G-C base pair is converted into a G-U base pair. When this mutated DNA is replicated, one strand will still have a G-C pair and is normal. However, the base pair in the other strand is made into a U-A pair because U is similar to T. The result of the nitrous acid damage then is to convert a G-C base pair to a T-A base pair. A chemical in the second category is caffeine, the stimulant in coffee and cola drinks. Caffeine's structure is similar to that of a nucleotide, but it is not exactly the same. Thus, when caffeine is incorporated into the structure of DNA, its base sequence can become altered after replication.

How Do We Detect Mutagens?

Now you may be a bit alarmed to hear that common chemicals found in food and beverages can be mutagenic. But how much of a danger to humans are mutagens? How can we measure their risk? There are a variety of methods used to determine whether some compounds are mutagenic. Each method has its own advantages and disadvantages, and often a combination of methods is necessary to determine risk, in particular for humans.

One method that is relatively cheap and easy to use is to test a potential mutagen on bacteria. After all, they have DNA just like us, but they are much smaller and easy to grow in large numbers in the laboratory. The test for mutagenic activity in bacteria is called the Ames test, named after Bruce Ames, its inventor. This test involves growing bacteria that are unable to grow in a particular nutrient broth unless they are mutated. Then, the number of bacterial colonies that grow after exposure to mutagens is compared to that without exposure to mutagens. This ratio gives us the mutagenic effect of the substance being tested. Although this test allows us to test substances on many bacterial cells and is inexpensive to perform, it is not perfect because bacterial cells do not have all the complexities of animals. For example, caffeine is indeed a mutagen in the Ames test, but in humans, caffeine is quickly excreted and so does not have much of a chance to cause damage to our DNA. Taking the other example of a chemical mutagen mentioned above, nitrate or nitrite, both are used by bacteria as nutrients for growth. Because bacteria cannot convert nitrate or nitrite to nitrous acid, the actual mutagen, nitrate and nitrite are not mutagenic in the Ames test. So using bacteria to test the mutagenicity of caffeine and nitrates or nitrites gives us totally misleading answers regarding their effects on humans. Still, the Ames test demonstrated that certain chemicals, some formerly used in hair dyes, for example, are mutagenic. This result prompted a ban on these chemicals for human use.

A second method involves testing potential mutagens in animals. As you can imagine, this is much more expensive. However, an example of a chemical identified as highly mutagenic by this method is vinyl chloride, a substance used to make polyvinyl chloride (PVC), a plastic used in plumbing. This chemical was originally used in beverage containers! Before testing this chemical on rats, the worker exposure limit set by the Occupational Safety and Health Administration (OSHA) was 500 ppm (parts per million) in the air over an eight-hour period. Then it was discovered that vinyl chloride causes cancer in rats at 50 ppm. Subsequently, in 1974 the worker exposure limit was dropped to 50 ppm and, the next year, down to 1 ppm. Vinyl chloride is now completely banned from use in beverage containers. Workers who were exposed to vinyl chloride prior to this new information are closely monitored for cancer. If animal tests were not done, it would have taken much longer to find out the dangers of this chemical.

Unfortunately, some chemicals are not recognized to be mutagenic until it is noticed that unusually high numbers of people in a community, a work place, or an occupation are afflicted with rare cancers or have a high number of miscarriages. This was the case with asbestos. We now know that there is no safe level for tiny asbestos fibers that can enter our lungs. It took some time to determine that asbestos was mutagenic because decades passed between exposure and the time when disease symptoms first appeared. Today, asbestos is associated with lung cancer, which is made worse by smoking, and a rare cancer of the lining of the chest and abdomen. Millions of workers who did not know the dangers of asbestos and worked in the manufacturing and repair of brake and clutch assemblies, as well as those involved in insulation work, are now suffering from cancer.

Why do we look at cases of cancer and miscarriages to determine whether a chemical is mutagenic or not? In an adult body, there are trillions of cells. When a mutagen alters the DNA in one single cell, even if the damage is very serious and the cell dies, this effect will not be noticeable. However, if damage occurs in one of our reproductive cells, a sperm or egg, a future individual may be affected. Indeed, the damaged sperm or egg may produce a mutated individual or, if the damage to the reproductive cell is too severe, an embryo may not even develop very far and may spontaneously abort. However, damage occurring in a single cell of an adult body can sometimes lead to cancer. When an affected cell does not die but instead starts dividing uncontrollably, it is cancerous. That is why rates of cancer, birth defects, and spontaneous abortions are used to try to assess exposure to mutagenic agents.

Teratogens

Mutagenic agents are not the only source of birth defects and spontaneous abortions. Some substances are more properly classified as teratogens instead of mutagens. Teratogens are substances that cause malformation of the developing embryo but do not act through changes in the DNA. A helpful analogy to understand the difference between teratogens and mutagens is to consider a building and its blueprint. A mutagen changes the DNA blueprint for an organism. The building or organism is flawed because the originally correct blueprint is altered. A teratogen does not alter the blueprint, but rather causes a defect in the building by providing faulty materials or making mistakes in the building process. Both teratogens and mutagens may cause spontaneous abortion, severe malformation of the embryo, or mental retardation of the baby, but the underlying mechanisms are very different.

Why do substances, some of which do not cause any problems in an adult nor cause changes in the DNA, cause major defects in a developing embryo? Early developmental processes have an immense effect on the individual, both mentally and physically. Remember that each of us begins as a fertilized egg that must divide many times in order to make all the different cells of our body. In order to develop properly, cells must interact with neighboring cells, and some must even move over other cells to be accurately positioned. These cellular activities are very prominent early in embryonic development. Any untoward influence on cells during these critical early stages will affect the processes in progress and a great percentage of cells in the developing embryo.

An example of a tragic teratogen is thalidomide, a medication formerly prescribed for nausea. This medication is perfectly safe for adults, and so it was prescribed for morning sickness to pregnant women in late 1950s and early 1960s. It was only after a spurt of babies were born with deformed limbs that anyone realized that thalidomide is a strong teratogen. It turns out that the teratogenic effect occurs in the very earliest days of pregnancy, when limbs are just developing and when the mother may not even be aware that she is pregnant. It is now thought that thalidomide affects the formation of blood vessels, a critical process in early development. Because of this effect, thalidomide is now being considered for cancer treatment, since tumors also need a good blood supply to grow.

Box 8.1 *Why There Were Few Thalidomide-Caused Birth Defects in the United States*

We now know the tragic consequences of taking the drug thalidomide for pregnant women. However, when this drug was first submitted for use in the United States as sedative to the U.S. Food and Drug Administration (FDA) in the early 1960s, little information was available to suggest its potential dangers. The story of how a tragedy was barely averted in this country speaks well of drug-safety laws in the United States and exemplifies the efforts of one woman who fought pressures to allow thalidomide to be sold in the country.

Thalidomide was first made in Germany in 1953, and by the late 1950s it was widely used as a sleeping pill by adults and children. It also prevented common morning sickness due to pregnancy and was widely used for that purpose. In 1960, Merrel Company applied to the FDA to market thalidomide in the United States. By this time, laws in the States required that drugs be demonstrated to be both safe and effective. Years of thalidomide use in Europe had not detected any toxic effects of the drug. Thus many felt that it would

Box 8.1 *continued*

be routinely approved in the United States. However, Dr. Frances Kelsey, newly hired by the FDA, carefully reviewed the application and noted several omissions that prevented her from judging the safety and efficacy of the drug. Merrel Company was eager to start tapping the lucrative U.S. market and, assuming that thalidomide would be approved, had brought in five tons to warehouses in the country. This company had even distributed the drug as an investigational drug to over a thousand doctors, a practice that is illegal today but was legal in 1960. In the meantime, Dr. Kelsey found out that there were cases of damage to nerves in the hands and feet with long term use and potential danger to the fetus. Merrel Company, eager to get approval, suggested releasing the drug with a warning label. Fortunately, Dr. Kelsey resisted all the pressure and requested more information on the drug's safety. By late 1961, birth defects of children born to mothers who had used thalidomide early in their pregnancy was discovered in Germany, and thalidomide was withdrawn from the market.

Tens of thousands of babies were born with terrible malformations in Europe. These malformations typically affect the limbs, with some having hands and feet sticking straight out of the torso without arms and legs. In the United States, only about ten babies, born to mothers given thalidomide as an investigational drug, were born with these defects.

Other teratogens have less specific effects. Alcohol is also a teratogen, though its effects are much less specific than that of thalidomide. Babies born to mothers who drink excessively are born with mental deficiencies and other defects. A final example of teratogen is diethylstilbestrol, also called DES. In the early 1950s it was believed that insufficient amounts of estrogen in pregnant women were causing miscarriages. In order to prevent this problem, DES, a substitute for estrogen, was routinely prescribed. This was done despite the fact that there was no medical evidence showing that DES reduced miscarriage rates. Only much later was it discovered that females exposed as fetuses to DES in the womb had a higher chance of developing problems with reproduction. These problems include cancer of the reproductive tract, infertility, and miscarriages. Why did DES preferentially affect female offspring? The difference in gamete development between the two sexes explains this, as we will see next.

Human Reproduction

There is a tremendous difference between men and women in the way they form their reproductive cells. First, let us begin with a short description of spermatogenesis, the process of making sperm. Spermatogonia (sperm stem cells) are the first cells in the lineage that leads to sperm. They have the ability to continually divide by mitosis. At puberty, some of these spermatogonia begin the process of meiosis, and each produces four haploid cells containing twenty-three chromosomes. These haploid cells also begin the process of compacting the DNA in their nucleus, forming the sperm head, and gradually developing the sperm tail, with a concomitant loss of much of their cellular material. This process, from beginning to end, takes almost fifty days. It is a continuous process, so once it begins, the testes contain reproductive cells at every stage of development. Mature sperm cells accumulate, and a single ejaculate contains hundreds of millions of sperm.

In quite a contrast, the process of making ova or eggs in the female begins before birth. While the young female embryo is still barely recognized as a pregnancy, the egg stem cells divide to make oocytes. By the eighth week of fetal development, the total number of oocytes that a female will ever carry is present, about half a million. These oocytes then begin the process of meiosis. Unlike sperm, however, oocytes do not divide equally in meiosis. Only one out of the four cells with a haploid chromosome number becomes a potential egg. But even further removed from spermatogenesis, meiosis in oocytes stops after the DNA replicates. This happens well before the female baby is born. Thus, all half-million potential egg cells in a female are in "suspended animation" at this early stage of meiosis. At puberty, one or two eggs are released each month in ovulation. Upon ovulation, the first of two divisions of meiosis occurs in the released egg. This means that some eggs are in this suspended state of animation for over fifty years! Then, only if a sperm fertilizes this egg does the egg quickly complete the second division of meiosis to make the haploid egg. In this way, the new individual is result of the joining of a haploid egg and a haploid sperm.

The above process of forming our gametic cells is important to understand, especially in light of the action of mutagens and teratogens. When we are exposed to mutagens or teratogens, our potential progeny, our reproductive cells, are also getting exposed. In the case of males, the cells that go on to make sperm are vulnerable to mutagenic agents. Because sperm cells are continually being made, sperm cells were thought to be less vulnerable than female gametes, which are never renewed. However there are now documented cases of male exposure to mutagens that result in higher incidence of leukemia among their children. Of course, for females, the window of vulnerability begins even before birth. Thus, exposing unborn females to radioactive fallout has had long-term consequences for girls born to exposed mothers.

Summary

In this chapter, we learned that mutations naturally occur during the metabolic processes of life as well as in DNA replication, which is not perfect. The rate of mutation above this natural low spontaneous rate can be increased by mutagens, whether they are forms of electromagnetic radiation or chemical mutagens. There are different methods for determining whether a substance is mutagenic or not. However, some substances are not discovered to be mutagenic until they have harmed people. Some mutagenic substances have been detected by monitoring cancer rates and rates of spontaneous abortions and birth defects. Teratogens are substances that are not mutagens and do not harm adults but can cause birth defects. The differences in the formation of gametes among males and females affect the window of vulnerability of our reproductive cells to mutagens.

CHAPTER 9

Linkage and Mapping
Gene Discovery

AS WAS DISCUSSED IN CHAPTERS 5 AND 6 and will later be talked about in chapter 13, bacteria, plants, and animals have been genetically modified. This means that donor genes first had to be identified. Also, to properly diagnose diseases, it is necessary to pinpoint the defective gene. Identifying defective genes may then help with conventional therapy, as well as gene therapy. In order to accomplish these goals, genes first must be located, that is, their location on chromosomes be determined and their sequences deciphered.

We now know that higher organisms contain many thousands of genes. For example, the fruit fly harbors 16,000 genes, while humans have about 35,000. As a step in the identification of a gene, we need to determine the position of these genes in relation to other genes on the chromosome. We will now see how it is possible to map genes on chromosomes, thus locating and isolating genes of interest.

There Are Many Genes on Each Chromosome
Chapters 2 and 3 described the inheritance pattern of genes located on the X chromosome. As we saw, it is the fact that sons always get their X chromosome from their mothers that allows the localization of several genes to that chromosome. Remember that we studied three such

genes, the genes for hemophilia and color blindness in humans, as well as the gene for eye color in fruit flies. Very soon after Morgan discovered the behavior of sex-linked genes in fruit flies, he found that other genes were also located on the X chromosome. These included genes determining body color and wing size. Clearly, several genes are located on the X chromosome. This holds true for all the other chromosomes. Also, since the fruit fly has four pairs of chromosomes and 16,000 genes, one can calculate that each chromosome must carry thousands of genes. Humans have twenty-three pairs of chromosomes. Given that we have about 35,000 genes, the average number of genes per chromosome is about 1,500. Of course, long chromosomes are expected to carry many more genes than short ones.

Independent Assortment of Genes

Let us now consider *two* genes, located on different chromosomes, and see how these genes behave in a cross. This situation is a bit more complex than crosses involving a single gene existing in two forms, as we studied in chapters 2 and 3. To illustrate what happens, let us use traits that we mentioned already.

For example, let us see what genotypes of offspring a man heterozygous for sickle-cell anemia, who is also carrier of the galactosemia gene, can have with a woman who is also heterozygous for sickle-cell anemia and a carrier for galactosemia. Remember that both conditions are recessive so both parents are phenotypically normal. Let us call **B** the normal β-hemoglobin gene, and **b** its sickle cell counterpart. Further, let us use **g** for the galactosemia gene and **G** for the normal gene. The genotypes of both parents are thus **BbGg**. They will not show any sign of disease since they are heterozygous for both traits. We learned in chapter 2 that single sets of chromosomes are present in gametes. It is also known that the genes for sickle-cell anemia and galactosemia are located on different chromosomes. What kinds of gametes will these parents produce?

Each gamete has one set of chromosomes. So each gamete will have one copy of the hemoglobin gene and one copy of the "galactosemia gene," either in its normal or abnormal form. Thus we can observe gametes with the gene combinations: **BG**, **Bg**, **bG**, and **bg**. These genes follow the third law of genetics, *the law of independent assortment*. This law states that genes located on different chromosomes assort independently. In other words, the B gene does not preferentially associate

A

Chromosomal composition
of parents

B

Chromosomal composition
of gametes

C

All possible combinations
of chromosomes
of offspring

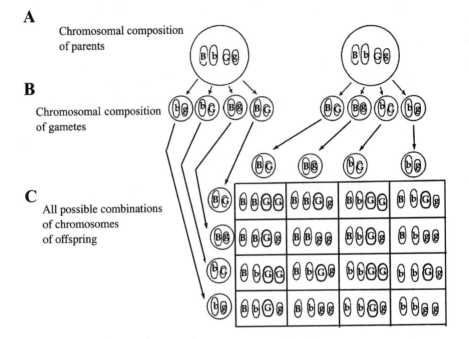

Figure 9.1 Independent Assortment of Two Pairs of Genes. The sixteen possible gene combinations observed in a cross involving two different pairs of genes located on separate chromosomes. A. The genotypes of the parents are shown with the genes located on separate chromosomes. Only the two pairs of chromosomes are shown for simplicity. B. The gene combinations of gametes are shown. C. A Punnett square shows all sixteen possible gene combinations of the cross.

with the **G** gene. Rather, **B** can just as easily associate with **g** as **G**. In our example, the mother and the father will produce gametes having the same four chromosomal compositions; this is because they are both heterozygous for both traits. We can now build a Punnett square to see what kinds of offspring these two parents can have. In this case, since we have four possible gametes for each parent, the Punnett square will contain sixteen boxes. Remember that if only two contrasting genes are considered, the Punnett square contains only four boxes. Figure 9.1 shows that when *two pairs* of contrasting genes are studied, the situation becomes more complicated. Note that some gene combinations (genotypes) are represented more than once in this Punnett square. By careful counting, we see that nine different genotypes are possible (see table 9.1). Then, taking into account that **B** is dominant over **b** and that **G** is dominant over **g**, we can calculate the probability

Table 9.1 All Possible Genotype Combinations Produced by Parents with Genotype **BbGg**

Genotype	Number	Phenotype	Total Number of Different Phenotypes
BBGG	1	normal	
BBGg	2	normal	
BbGG	2	normal	
BbGg	4	normal	total normal = 9
Bbgg	2	galactosemic	
BBgg	1	galactosemic	total galactosemic = 3
bbGG	1	sickle cell	
bbGg	2	sickle cell	total sickle cell = 3
bbgg	1	sickle cell & galactosemic	total both = 1

of individuals with various phenotypes. The individuals who are **BBGG, BBGg, BbGG** and **BbGg** will have the same phenotype: no symptoms of sickle-cell anemia and no symptoms of galactosemia. Next, the individuals who are **BBgg** and **Bbgg** will show no signs of sickle-cell anemia, but will be galactosemic. Then, those who are **bbGG** and **bbGg** will have sickle-cell anemia, but no galactosemia. Finally, the individuals with the **bbgg** genotype will unfortunately suffer from both diseases. Thus, crosses involving two pairs of contrasting genes produce nine different genotypic categories and four different phenotypic categories.

By carefully counting the boxes in the Punnett square that contain different genotypes, we realize that the four phenotypic categories come in a 9:3:3:1 ratio. Nine out of sixteen boxes (~56 percent) represent the probability of being phenotypically normal for both traits. Three out of sixteen boxes (~19 percent) fall in the category where the first phenotype is normal (no sickle-cell anemia) but the other is not (galactosemia is present). Next, three out of sixteen boxes (~19 percent) represent the probability for an offspring to have sickle-cell anemia but not galactosemia. Finally, there is a 1 in 16 probability (~6 percent) that double heterozygous parents will have an offspring with both sickle-cell anemia and galactosemia.

This is different from a cross involving single genes (existing in two forms), where only three possible genotypes and two phenotypic categories are observed (see chapter 2). There, remember that the phenotypic ratio is 3:1. As you can well imagine, things get even more complicated if three pairs of genes are considered. In this case, the

Punnett square contains sixty-four boxes, and there are eight possible phenotypic classes. In the human case, forty-six chromosomes (twenty-three pairs) segregate and assort independently during meiosis. This means that 2^{23} possible phenotypic classes are potentially generated from each offspring, simply from the independent assortment of chromosomes into the gametes! You can now understand why each human is unique. Identical twins are an exception to this rule because they originate from the splitting of a single fertilized egg. For an example of independent assortment, see the "Try This at Home" at the end of this chapter.

Linkage

We have already used the term "linked" to refer to genes located on the sex chromosome. These genes are called "sex-linked." We saw at the beginning of this chapter that, given the large number of genes in an individual and given the limited number of chromosomes present in that individual's cells, each chromosome must carry a large number of genes. When genes are located on the same chromosome, they are said to be linked. In the example above, which uses sickle-cell anemia and galactosemia as a pair of contrasting traits, the genes are on different chromosomes and are thus unlinked.

Gene linkage was elucidated in the laboratory of Thomas Hunt Morgan at Columbia University. As we explained before, Morgan generated many types of mutant fruit flies and crossed them. One cross involved flies with mutant wings, symbolized by **r** (the normal wing is represented by **R**), and black body indicated as **b** (**B** indicates again the dominant, normal body-color gene). Since mutant wings and black bodies are both recessive, these flies must be homozygous for the trait, and their genotype is thus **rrbb**. They were crossed with normal flies of genotype **RRBB**. Remember that normal is dominant over mutant. All the offspring of this cross are double heterozygotes, **RrBb**, and thus are phenotypically normal. Then, Morgan crossed these double heterozygous offspring. This cross, written **RrBb** x **RrBb**, involves two different pairs of traits, equivalent to the example with sickle-cell anemia and galactosemia in humans that we saw above.

Thus, in the offspring of his cross, Morgan should have observed four phenotypic classes: normal wings and body color, normal wings and black bodies, mutant wings with normal body color, and mutant wings with black bodies. Furthermore, these four categories should be produced in 9:3:3:1 proportions (56 percent, 19 percent, 19 percent, 6 percent) following the rule of independent assortment. Morgan did observe four categories, but the numbers of flies in each category were completely off! He observed 74 percent normal, only 1.5 percent each of just mutant wings or just black body, and 22 percent that had both mutant wings and black body. That is, he observed many more totally normal flies as well as flies with mutant wings and black body (**rrbb**) than expected. Further, he observed extremely low numbers of flies that had normal wings with a black body, or mutant wings with normal body color.

What can account for this gross deviation from the expected proportions of different phenotypes that we expect from the rules of independent assortment? Morgan correctly interpreted these results: he hypothesized that the **R** and **B** genes were on the *same* chromosome from one of the original parents. Similarly, the **r** and **b** genes were also located on the same chromosome, this time the one from the other original parent (figure 9.2). It is because the genes Morgan studied were *not* assorting independently (i.e., were *not* on different chromosomes) that a 9:3:3:1 ratio was *not* observed. Genes that do not assort independently must be on the same chromosome; they are linked.

If the two genes are on the same chromosome, figure 9.2 predicts that we should have gotten 75 percent flies with normal body color and normal wings (**RRBB** or **RrBb**), and 25 percent flies with mutant wings and black body (**rrbb**). We should *not* have gotten *any* flies that had just black bodies or mutant wings. So, how can we account for the small but significant numbers of flies that had just black bodies or mutant wings? The reason is that chromosomes can break and rejoin and, in the process, shuffle genes. This process is called "recombination."

Recombination

During meiosis, chromosomes come into close physical contact. This can lead to chromosome breakage and rejoining. In this process, an overlapping chromosome breaks at the point of contact with another chromosome, and the two ends of the broken chromosomes can swap positions. Usually, chromosomes break and join at the same position so there is no gain or loss of genes in this process. But the result is that genes present on different copies of the same chromosomes can

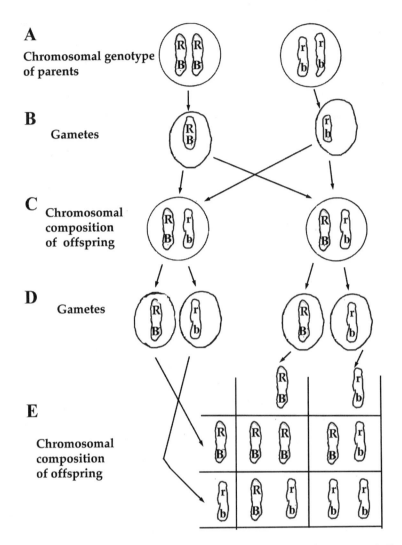

A
Chromosomal genotype
of parents

B
Gametes

C Chromosomal
composition
of offspring

D Gametes

E
Chromosomal
composition
of offspring

Figure 9.2 Inheritance of Two Pairs of Genes Located on the Same Chromosome. A. The genotype of double-homozygous parents are shown with both the eye color genes (**R** and **r**) and wing gene (**B** and **b**) on the same chromosome. B. Each parent produces only one type of gamete since they are homozygous. C. The offspring of the cross produces a double heterozygote, but note that the dominant form of the two genes are on the same chromosome and the recessive form of the two genes are also on the same chromosome. D. The double heterozygote offspring in this linked case produces only two types of gametes, not four types as seen in figure 9.1. E. There are only three different genotypes of offspring resulting from the double heterozygote parents if the genes are tightly linked on the chromosome.

get shuffled. Let's take the above example of the mutant wings and black bodies of fruit flies. First, figure 9.2 shows the result in a situation where chromosomes do not break and rejoin. Here, the **R-B** and **r-b** genes stay together and only two phenotypic classes are observed: 75 percent of the flies have normal wings and normal body color, whereas 25 percent of the flies have mutant wings and black body color. No recombinants are seen. Morgan did not obtain these numbers because he did observe recombinants. Let us now imagine that these chromosomes, one with the **R** and **B** genes, and one with the **r** and **b** genes, break and rejoin between **R** and **B** and between **r** and **b** (figure 9.3). You can see that the **R** gene, originally linked to the **B** gene, is now associated with the **b** gene. Similarly, **r** is now associated with **B**. These new gene combinations are due to chromosome recombination and the resultant chromosomes are called recombinants. Individuals with the recombinant chromosomes are also called recombinants.

How often does recombination occur? It is the distance between genes along the chromosome that determines how often they recombine. You can imagine that breakage anywhere between the **R/r** and **B/b** genes will result in the new combination shown in figure 9.3. Thus the frequency of recombination depends roughly on the distance between **R/r** and **B/b** genes. For example, if the two genes are at the opposite ends of the chromosome, the chances of recombination are high. In that case, there will be many recombinants. If, on the contrary, the two genes are very close together, chances of recombination between the two genes are small. In Morgan's experiment, the distance between the two genes, **R/r** and **B/b**, was pretty close: he observed only 3 percent recombinants.

By determining the percentage recombination of linked genes in the offspring of a cross, we can map genes onto chromosomes. With fruit flies, if appropriate crosses are made, scientists can decide whether the genes are unlinked. This is the case when they observe a phenotypic ratio expected for independent assortment, for example a 9:3:3:1 ratio. If the ratio is significantly different from this expected ratio, they can conclude that the two genes are on the same chromosome, linked. Next, if two genes are linked, the proportion of individuals with phenotypes resulting from recombination is a measure of how far apart the genes are located on the chromosome. In fact, a standard unit of distances along chromosomes is the recombination

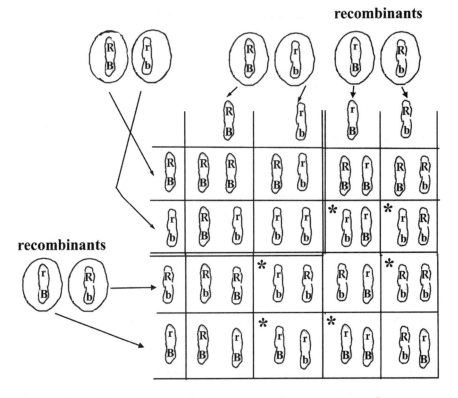

Figure 9.3 Breaking and Rejoining the Chromosomes Shown in Figure 9.2. The Punnett square shown in figure 9.2.E is expanded to include the chromosomes resulting from recombination. The original Punnett square is shown on the upper left enclosed by double lines. The recombinant chromosomes are added to the Punnett square. Those boxes marked by asterisk (*) represent the phenotypic classes not seen in the figure 9.2. Because the recombinant chromosomes represent only 6 percent of the chromosomes, the numbers of flies with the * phenotype are small.

percentage, named centimorgans after T. H. Morgan. We can thus establish a chromosome map giving the relative positions of many genes, their distances measured as the amount of recombination observed between them.

We see that genes can be mapped on the chromosomes of *Drosophila* because we can do many crosses between flies with different genetic mutations. Then we can analyze hundreds of their offspring to see if the numbers correspond to ratios for independent assortment or not. In humans, we cannot do this. How do we measure linkage in humans?

Linkage to a DNA Marker

Since gene mapping with crosses by linkage is generally not feasible in humans, how do we map human genes? One technique used to find the gene corresponding to a disease is called restriction fragment length polymorphism, or RFLP for short. The first word, restriction, refers to the first step in the process, which is to cut the DNA with restriction enzymes (see chapter 5). Let's consider what happens to our DNA when we cut it with a restriction enzyme. A helpful analogy is to imagine a large document like an encyclopedia, though printed on a continuous roll of paper. Now, our "restriction enzyme" would be like using scissors to cut every place where there was a particular sequence of letters, for example the sequence of letters in "tem." Then we would cut at words like *tem*porary, *stem*, *system*, and so on. How many and what sizes of pieces do we get? In some articles of the encyclopedia, there may be many small pieces, but there may be long regions with no cuts and a wide range of sizes in between the small and the large pieces. With DNA, we would also get a range of sizes of fragments from small to large when we cut using a restriction enzyme that cuts DNA at a particular base sequence. So this is what FL in RFLP refers to, the fragment lengths created by cutting our DNA with restriction enzymes. Would a given restriction enzyme cut everybody's DNA at always the same place and thus produce DNA pieces of the same size? The answer is no because individuals have differences in their DNA sequence. Therefore, when one cuts DNA with restriction enzymes, different individuals, even close relatives, produce different sizes of DNA pieces. The word polymorphism means "many forms," in this case, many DNA fragment lengths. Thus we say that the lengths of our DNA fragments are polymorphic. But our genome contains billions of nucleotide base pairs, so when we cut our DNA with a restriction enzyme, there are hundreds of thousands of different sizes of DNA. If we now separate these DNA fragments using gel electrophoresis (recall that we used gel electrophoresis to separate DNA by size in chapter 1), we would not see bands of DNA of specific sizes but a smear of DNA running from the largest to the smallest fragment lengths. So even though different individuals have their DNA cut at different positions, because the DNA sequence is a little different between even closely related individuals, we would not be able to distinguish them by analyzing the cut DNA with gel electrophoresis alone.

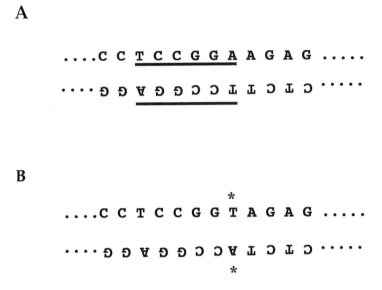

Figure 9.4 A Portion of the β-hemoglobin Gene with Codes for Normal and Sickle-cell Traits. A. This portion of the normal hemoglobin gene contains a restriction enzyme site, which is underlined. B. The same region shown in A for the hemoglobin of an individual with sickle-cell anemia. The asterisk shows the single nucleotide mutation. This mutation makes the restriction enzymes unable to recognize the site.

So how do we tell the difference between individuals? Remember from chapter 1 that DNA is a double helix held together by complementary base pairing. Recall also that we can now make single-stranded DNA molecules of a specific base sequence in the test tube and use them as primers for polymerase chain reaction. We can use these tools, restriction enzymes and DNA with specific sequences, to distinguish between the DNA from two different individuals. Let's take the case of sickle-cell anemia. A short region of the DNA encoding normal β-hemoglobin is shown in figure 9.4.A. The underlined region indicates a restriction-enzyme site, a place that will be cut by a certain restriction enzyme. The mutation in sickle-cell anemia is a single nucleotide change indicated by an asterisk in figure 9.4.B. Because the base change occurs within the restriction enzyme site, the restriction enzyme no longer recognizes this sequence and so will not cut the mutant DNA at this position. The change in the cutting pattern of DNA is illustrated in figure 9.5.A. This is an example of a single-nucleotide polymorphism (abbreviated SNP) because only a single nucleotide was changed.

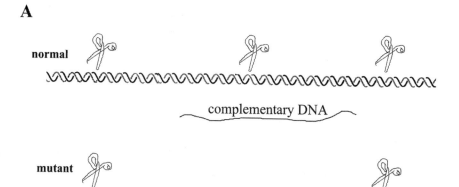

A

normal

complementary DNA

mutant

B

marker   sickle   normal   carrier

Figure 9.5 Restriction Fragment Length Polymorphism of Sickle-cell Anemia. A. Double strands of DNA are depicted with the positions of restriction enzyme cuts marked with scissors. The upper DNA shows a small portion of the normal β-hemoglobin gene from an individual with normal β-hemoglobin; below it is the same section from an individual with sickle-cell anemia. Note that a restriction site is now missing due to the mutation. The strand of complementary DNA is positioned between the two DNA double strands. B. A gel electrophoretic analysis of DNA from a sickle-cell anemia patient. The marker lane contains size markers; the sickle lane shows the long single fragment found in the mutant. The normal lane shows the two smaller bands of DNA corresponding to the two DNA fragments created by the restriction enzymes. The carrier lane, as expected, has bands representing both the mutant and the normal DNA.

Figure 9.5.B shows the gel electrophoresis pattern of several individuals, two of whom have at least one gene for sickle-cell anemia. The specific bands are differentiated from the background DNA smear because they can bind to the "complementary DNA" probe specific to a region of the β-hemoglobin gene. The complementary DNA can be labeled with a dye for easy detection. No other bands present in the DNA smear binds the complementary DNA, and so they remain invisible. Another example of this technique is shown in box 9.1.

In the sickle-cell anemia example, the defect is a specific change in the DNA sequence of a specific gene , the β-hemoglobin gene. This means that the band of DNA in RFLP is *always* associated with a sickle-cell defect. However, most often we do not know what gene or what defect in a gene is associated with a disease.

Box 9.1 *Identifying Disease Genes Using Restriction Fragment Length Polymorphism*

Restriction fragment length polymorphism allows us to identify the piece of DNA associated with a genetic disease. In this procedure, the DNA of affected and unaffected individuals are cut with the same restriction enzyme. If we arranged these pieces by size, we would get a large range, from small to large, and there would not be a distinct group of pieces of a particular size. Recall from chapter 1 that gel electrophoresis is used to separate pieces of DNA by size: if too many pieces of many different sizes are run together, the resulting gel would show the DNA spread out in a smear.

So how can we tell where a particular piece of DNA is in this smear? Recall that DNA has a double-helical structure held together by complementary base pairs. We can make a short piece of DNA with a specific sequence and label it with a dye. After the DNA is cut and put on a gel and separated by size, we can transfer these pieces of cut DNA onto a sheet that prevents them from floating away. By using appropriate chemicals, we can then separate the double strands of the cut DNA pieces. At that point, the specific short piece of labeled DNA is added and allowed to bind to the complementary sequence present somewhere among the cut pieces of DNA on the sheet. Recall from chapter 1 that short pieces of DNA find the complementary partner faster than large pieces.

Box 9.1 *continued*

After that, we wash away any labeled DNA that has not bound to its complementary sequence. The band identified by the labeled DNA is a piece of DNA that has within its length a sequence complementary to the labeled DNA. The position of the labeled band determines the size of the DNA piece and how far it moved in the original gel electrophoresis.

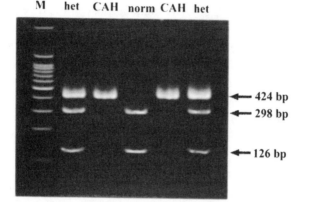

Figure B.9.1 RFLP to Identify Congenital Adrenal Hyperplasia. Gel electrophoresis of DNA from normal individuals (norm), those that are heterozygous (het) for congenital adrenal hyperplasia, and congenital adrenal hyperplasia patients (CAH). M provides size markers used to determine the size of the bands shown on the right as numbers of base pairs (bp). Note that both affected individuals (CAH) have a single band at 424 base pairs whereas normal individuals are missing that band and have instead two bands at 298 base pairs and 126 base pairs.

Now, we can treat this band of DNA on the gel as a genetic trait in the same way we view a person's blood type. Figure B.9.1 shows an example of identifying the genetic defect for congenital adrenal hyperplasia. Congenital adrenal hyperplasia is a recessive disease caused by a defect in an enzyme important for salt balance. Infants with this disease can develop life-threatening dehydration or shock. This disease is treatable with hormones.

Even in cases where the disease mutation itself does not produce different banding patterns in RFLP, this technique can help track down the gene responsible for a disease. This is because a special banding pattern may be associated with a disease. Why might this be? This would be the case if a particular DNA sequence is closely linked to the disease gene, that is, located close to the disease gene on the chromosome. Recall that recombination between genes is caused by the breaking and rejoining of chromosomes. The recombination frequency depends upon the distance between the two genes. So we can consider a band on a gel as a gene. If the rate of recombination between a particular DNA sequence and the actual defective gene that causes the disease is low, the distance between them must be very short, that is, that piece of DNA must be close to the disease gene. By this logic, one can find a fragment of DNA that is close to the defective gene. The closer the piece of DNA represented by the restriction fragment, the more likely that piece will be inherited together with the defective gene. Since the restriction fragment can be detected by gel electrophoresis, we can locate the defective gene closely associated with it. Researchers analyze the DNA of large families in which the disease exists; affected family members' DNA should show a higher chance of being associated with a specific DNA sequence close to the disease gene, whereas those of unaffected members should not. It is by this technique that scientists have closed in on genes responsible for many human genetic diseases.

Box 9.2 *Identifying a Disease-Resistance Gene in Barley Through Map-Based Cloning*

The most interesting and important genes, such as disease-causing genes in humans or disease-resistance genes in plants, are often the most difficult to identify because we do not know what proteins they code for or what their function might be. Cloning a gene is the first step toward learning how it functions and using it to suit our needs.

Stem rust is a fungal disease in barley that routinely reduced crop yields until the early 1940s. In a particularly bad epidemic of the disease in 1935, Sam Lykken, a farmer from Kindred, North Dakota, identified a single healthy plant in his field, which was otherwise totally decimated by stem rust. This single healthy plant was saved and became the source of resistance genes in barley strains. It turns out that this strain of barley contained the dominant stem

Box 9.2 *continued*

rust–resistance gene that has since been given the name "Rpg1." The presence of this gene has prevented any significant losses due to this disease since then.

To learn more about the stem rust–resistance gene in barley, a team of scientists led by Andris Kleinhofs at Washington State University in Pullman, Washington, set out to clone this gene. How can one identify this gene or other genes that are known only by their phenotype? We can look for DNA markers or bands on gels linked to the trait. The barley genome is huge; thus a gene can be hundreds of thousands of base pairs away from the DNA marker to which it is linked! So the scientists put large pieces of the barley genome into bacteria in such a way that each bacterium contained a portion of the barley genome. Recombinant DNA molecules like these are called bacterial artificial chromosomes, or BAC clones. One can identify the BAC clone of interest with the DNA markers linked to Rpg1. Because the BAC clones contain random pieces of the genome, if one has enough of them overlapping pieces can be found. Chromosome walking is a process used to follow the sequence of DNA in overlapping pieces of DNA like BAC clones. Once a number of overlapping DNA clones including the gene of interest are found, they are sequenced to identify potential genes and candidates for the gene of interest (figure B.9.2.A)

We say candidate genes at this point in the process because we still do not know which gene is the one responsible for disease resistance. Searching among the candidate genes to identify the resistance gene is difficult. The DNA sequences must be compared between resistant and susceptible strains to identify which gene is consistently different between them. If a particular gene always shows a difference, this is strong evidence, but not proof, that the correct gene has been identified. Kleinhofs's team used this method to identify the Rpg1 gene, but they also transformed barley with the Rpg1 gene. If a gene is indeed the one that confers resistance to stem rust, the transformed barley should be resistant. Indeed the transformation of a susceptible barley strain with the Rpg1 gene conferred resistance to the stem-rust pathogen! Thus Rpg1 is truly the stem rust–resistance gene! Interestingly, the transgenic plants

continued on next page

Box 9.2 *continued*

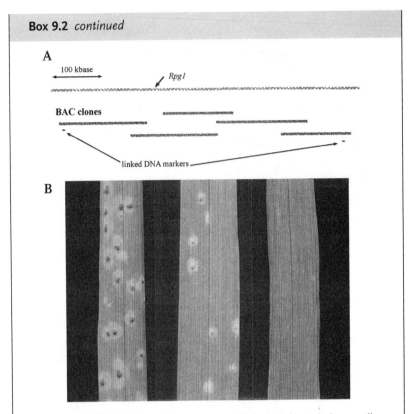

A

100 kbase

Rpg1

BAC clones

linked DNA markers

B

Figure B.9.2 Identifying the Disease-Resistance Gene in Barley. A. A diagram illustrating map-based cloning. Over half a million base pairs of the barley genome containing Rpg1, the stem rust–resistance gene, are shown at the top. The two DNA markers linked to Rpg1 are shown below the bacterial artificial chromosomes (BAC clones) that contain this DNA sequence. Additional contiguous BAC clones used to "walk" to the disease resistance gene are shown in dark mottled lines. B. Photographs of barley leaves treated with spores of the fungus that causes stem-rust disease. On the left is the susceptible strain, in the middle is the resistant strain, and on the right, a susceptible strain transformed with the resistance gene Rpg1. Note that the transformed strain shows no evidence of rust, and is thus more resistant than the original resistant strain. Photo courtesy David Hansen.

were more resistant to stem rust than the original resistant strain from which the gene was isolated (figure B.9.2.B). Though this was an arduous process that took years, the identification of the resistance gene by the Kleinhofs's team allows researchers to understand disease resistance better and to use this knowledge and this gene for agricultural purposes.

The Human Genome Project and Others

The procedure outlined above for mapping a disease gene on a chromosome is quite involved. Because of this, a major international scientific collaboration to sequence the whole human genome was set up. Sequencing the 3.15 billion base pairs of the human genome was accomplished in 2001. With this information in hand, finding and cloning the gene for a disease should be greatly facilitated. Indeed, after sequencing the DNA fragment associated with the disease gene, powerful software can scan the whole human DNA sequence and precisely position that fragment. By so doing, researchers can determine the sequence of the disease gene located next to the fragment used to probe the DNA.

DNA sequencing is largely automated today. First, DNA is isolated from an organism, broken into small fragments, and cloned in a plasmid vector. The sequences of these fragments are determined using chemicals that react with the four bases, and the sequences of the fragments are arranged in a linear fashion using powerful computer software. Bacterial genomes that consist of a few million base pairs can be assembled in a few months. Of course, it takes longer to sequence larger genomes. An interesting finding of the Human Genome Project was that most of the human DNA does *not* contain genes. Indeed, over 95 percent of human DNA does not code for proteins. This also holds true for other animals whose DNA has been sequenced. The function of noncoding DNA is not entirely clear. Much of it consists of simple sequences repeated thousands of times. It goes without saying that spotting human genes in a mass of DNA is not an easy task. This work is still in progress. To date, it is estimated that humans have about 35,000 genes, compared to 16,000 in the fruit fly and 19,000 in the simple nematode worm *Caenorhabditis elegans*.

Other genome projects aim at sequencing the fruit fly, mouse, rat, and chimpanzee genomes. Examples of plant genomes sequenced or with sequencing in progress are rice and *Arabidopsis*, a small plant with a small genome. Genomes of medical importance, such as that of the malaria parasite have been sequenced. Comparison of genes sequenced in humans and fruit flies have identified hundreds of genes that are so similar between them that scientists can use fruit flies to investigate genes implicated in human genetic diseases. Mice and rats are often used as model systems for the study of human diseases. Therefore, knowing their DNA sequence will facilitate the study of disease genes in animal models. Knowledge of the chimpanzee genome may tell us what genes make us different from our closest cousins.

Discovering Disease Genes in Humans

One of the goals of the Human Genome Project is to determine which genes make us prone to disease. This does not mean the types of genetic diseases that are well understood (such as PKU and others) that we described in chapter 3. Rather, we are referring to genes that make some of us more susceptible to certain diseases, such as bacterial infection, high blood pressure, and cancer. These diseases and many others involve many genes as well as environmental factors.

Now that the human genome has been deciphered, it should be possible to identify disease-susceptibility genes and attribute a function to them. This process is called "annotation." It is not enough to know the 3.15-billion base-pair sequence of the human genome if the genes contained in this sequence cannot be identified. Annotation is not an easy task, especially when you know that more than 95 percent of the human genome does not contain any genes. How does one discover genes in all that sequenced DNA?

One way to go about this is to compare the human genome with other, similar sequenced genomes. Mice have been used for a long time in genetic research, and many mutants, including behavioral and neurological mutants, have been created. In many cases, these mutations have been mapped precisely, and the genes involved have been located. Thus, once the mouse genome is completed, researchers will be able to annotate human genes based on similar locations on chromosomes and similar genes in mice. However, there may be better model than the mouse for annotating human genes.

Quite recently, investigators have sequenced the genome of the puffer fish, *Fugu rubripes*. It turns out that the genome of this fish contains only 365 million base pairs, one-ninth the number found in humans. It also appears that *Fugu* has about the same number of genes as humans, because it does not contain the enormous amount of noncoding DNA that humans have. But, you might think, how could fish genes help annotate the human genome? Thus it might surprise you that over 1,000 human genes have already been identified by their similarity to the *Fugu* genes.

But what about disease genes? Here too, researchers are optimistic that the *Fugu* sequence will help them identify genes in humans. The puffer fish evolved a very long time before humans did. Therefore, its genes have had a chance to change a lot more than human genes. Thus, if one finds in *Fugu* a gene that closely corresponds to a gene in humans, this means that this particular gene may have a fundamental role in survival. In fact, some researchers think that the *Fugu* sequence will be more important in that respect than the mouse sequence. This is because mice and humans are more closely related than either is to *Fugu*. In this case, annotation will not be easy because humans and mice are evolutionarily too close to each other, and it will be more difficult to decide which genes in both mice and humans are critical for survival.

Summary

We have studied the techniques used by geneticists to determine whether genes are located on the same chromosome, that is, whether they are linked. Phenotypic ratios of offspring from crosses with linked genes differ from the expected ratios for unlinked genes. This is because genes on the same chromosome tend to be inherited together. However, chromosomes can break and rejoin, resulting in recombinant chromosomes. The percentage of offspring with recombinant chromosomes gives us a measure of the distance between linked genes on the chromosome. In humans, where extensive, controlled crosses cannot be performed, other techniques are needed to determine gene linkage. One of these techniques consists in determining the close association of RFLP DNA fragments with a diseased phenotype. The full sequencing of the human and other genomes will allow efficient identification and study of disease genes.

Try This at Home: Independent Assortment of Chromosomes and the Making of a Unique Individual

This game will allow you to directly see how the independent assortment of chromosomes results in new gene combinations. In order to make the game not take too long, we'll consider an organism with only three pairs of chromosomes. Remember we actually have twenty-three pairs of chromosomes.

For each pair of chromosomes, we'll have two different types that we'll call heads (H) and tails (T) since we'll flip a coin to determine which we pick. Thus we have chromosome 1 heads (1 H)

continued on next page

and chromosome 1 tails (1 T), chromosome 2 heads (2 H) and chromosome 2 tails (2 T), and chromosome 3 heads (3 H) and chromosome 3 tails (3 T). We begin with a father and a mother, each with three pairs of chromosomes. We will indicate the chromosomes from the father with f and the chromosome from the mother with m. So the chromosomal compositions of mother and father are:

Father: 1 f H and 1 f T, 2 f H and 2 f T, 3 f H and 3 f T
Mother: 1 m H and 1 m T, 2 m H and 2 m T, 3 m H and 3 m T

Each team of two has a "mother" and a "father," with each person representing either the father or the mother. To begin the game, each person flips a coin three times to determine which of the pairs of chromosome they contribute to the next generation. The first flip determines whether chromosome 1 H or T is contributed, the second flip determines which chromosome 2 is contributed, and so forth. Then the team writes down the chromosome composition of their "child." If done properly, you will see that there is one f and one m chromosome for each pair of chromosomes. Now compare the chromosome composition of your "child" with that of other teams. Are there any with the identical chromosome composition? What is the chance that we would get two children with the same chromosome composition?

There is an explosion of genetic information being accumulated in databases around the world. Much of the information is available to anyone with Internet access. We will introduce you to some of that huge amount of information in this exercise to help you get acquainted with the databases available for genetic studies.

In this exercise, you will first find a gene associated with a human genetic disease. Then we will search the DNA database for the gene. Finally, we'll use one of many tools available to analyze DNA sequences, called BLAST (Basic Local Alignment Search Tool). We will use the example of Marfan syndrome that we encountered in chapter 3, but we encourage you to follow along in this exercise with another disease or issue of interest to you.

First, open an Internet browser and go to the Entrez search and retrieval system of the National Center for Biotechnology Information of the National Library of Medicine at the National Institutes of Health site at http://www.ncbi.nlm.nih.gov/Entrez/. At this site you will see a link called PubMed, which links to the professional biomedical literature database, another labeled Nucleotide, which links to the DNA-sequence database GenBank, and many others.

We will first look up a genetic disease in OMIM, the Online Mendelian Inheritance in Man, by clicking on "OMIM." Then, in the space provided, type in a disease or an organ of the body that you are interested in or even a characteristic of interest like left-handedness. As an example, we will type in Marfan and click "Go," which gives us a listing of OMIM entry numbers. Each of these listings provides authoritative information about the disease. Genes can be identified with a set of upper case letters sometimes followed by a number. When we click on #154700, the listing gives information about Marfan syndrome. At the beginning it states that "all cases of the true Marfan syndrome appear to be due to mutation in the fibrillin-1 gene (134797)." Clicking on that number gives us the name for the gene and its acronym, FBN1. In the column on the left, the light orange box labeled "R" provides a link to the reference-gene sequence. The default display is a short summary of the sequence. If you want more detailed information, you can

continued on next page

choose GenBank from the pull down menu and then click "Display" to show the information. For our purpose, however, we want to get the simple DNA sequence, so choose the FASTA format from the same menu and click "Display." Because this gene is huge, for demonstration purposes, select the first few lines of the sequence and copy it.

Now we will use BLAST, the Basic Local Alignment Search Tool, to find other related sequences. To do this, go to the BLAST page at http://www.ncbi.nlm.nih.gov/BLAST/. You will see the many types of BLAST searches that can be performed. There is a BLAST tutorial page at http://www.ncbi.nlm.nih.gov/BLASTinfo/information3.html. For our example, we will use the Translated BLAST search. This search takes a DNA sequence and translates it into protein sequences. The program can search through the protein database (blastx) or through the DNA database that is translated into protein (tblastx). We will do the latter by clicking on that line and then using the first few lines of our FBN1 gene that we copied by pasting it into the box next to "search." We can leave everything else at the default settings and click on "BLAST." You should now get a box with your request ID and an estimate of the time it will take to do this search. Depending upon how crowded internet traffic is in general and at the NIH site in particular, and depending on the size of your search (which includes both the query DNA sequence length and the database that you are searching), the search can take from seconds to many minutes, and sometimes hours. You can shorten the time needed for your search by using a shorter DNA sequence for the query and limiting your search to a subset of organisms. There are millions of people using this search tool throughout the world, searching through an ever-increasing database. Thus, even the few minutes that a search may take is quite fast!

A BLAST search seeks to find DNA sequences that are similar to the query sequence. The results display the closest match, first in a color-coded graph that shows the region of similarity, ranging from red being the closest match to black being the least similar. The higher the "Score" value, the better the match. The "E "value provides a statistical measure of how likely it is to have found this match by random chance. The smaller this value, the more likely it is that the match is meaningful. This tool is commonly used when one first sequences a piece of DNA in order to find the closest match and help identify the gene sequenced. The E value can also find similar genes in closely related or even distantly related organisms. For our example, not surprisingly, the best matches to our Marfan gene piece are to human fibrillin genes. But further down the list, one finds *Bos taurus*, cow, *Sus domesticus*, pig, *Mus musculus*, mouse, and *Rattus norvicus*, rat genes. If one goes down further, we can even find that the fruit fly, *Drosophila melanogaster*, has a gene similar to human fibrillin, admittedly with less similarity than the mammals listed above.

CHAPTER 10

Genetics of Populations and Genetic Testing

SO FAR, WE HAVE DEALT WITH THE INHERITANCE of genes in *single* individuals. The genetic study of a collection of individuals must take into account variations in the population and reasons for those variations. The word "population" simply refers to large numbers of individuals able to breed: for example, human populations. The genetics of whole populations is useful for predicting the chances of genetic diseases occurring in large groups of individuals. It is also the basis for understanding differences in the frequency of traits or diseases in different ethnic groups.

Why Don't We Observe 3 to 1 Ratios of Dominant Versus Recessive Traits in Populations?

In chapter 3 we saw that the laws of Mendelian inheritance apply to people too. At first, many scientists did not easily accept this idea. The first case of a dominant trait discovered in humans, in 1903, was brachydactyly. This condition is characterized by shortened or fused fingers and toes. Based on the low frequency of this condition in the population at large, some scientists stated that Mendelian genetics did not apply to people. One wrote, "If Mendelian Laws of Genetics apply to people, and brachydactyly is a dominant trait, why don't 3 out of 4 people have

	B	b
B	BB	Bb
b	Bb	bb

Figure 10.1 A Punnett Square Used to Predict the Offspring of Heterozygous Brachydactyly Parents. With these parents we would expect 75 percent brachydactyly children and 25 percent normal since brachydactyly is a dominant trait. However, this prediction is true only if both parents are heterozygous for brachydactyly.

brachydactyly?" (Udny Yule, cited in G. H. Hardy, "Mendelian Proportions in a Mixed Population," *Science* n.s. 28 (1908): 49–50). You see, this scientist drew a Punnett square with two heterozygotes for brachydactyly as parents and saw that 75 percent of the offspring must have the brachydactyly trait (figure 10.1). Since 75 percent of the population doesn't have brachydactyly, he claimed that Mendelian laws of genetics do not apply to people! Can you see what is wrong with his logic? The prediction of 75 percent brachydactyly only applies if two heterozygotes for brachydactyly have children, in which case the prediction is absolutely correct. Since most people do not have brachydactyly, most parents are normal and do not expect to have any brachydactyly children. The flaw in the logic was pointed out by an English mathematician, Godfrey H. Hardy, and by Wilhelm Weinberg, a German physician. Both provided the correct explanation as to why a dominant trait does not appear three times more frequently than a recessive trait in a population. Thus, predictions regarding genotype frequencies in populations is called the Hardy-Weinberg law. This law is useful for calculating approximate frequencies of genes and genotypes in populations.

Predicting the Genotype of the Next Generation
Using the Punnett Square

The Hardy-Weinberg law is a direct extension of how genes are passed down from parents to offspring on an individual basis to the same process for a whole population. Because of this, we will again use a

Punnett square. Recall that in setting up the Punnett square, we put the genotype of the gametes from one parent on top and the other parent on one side. Then, we carry down the gene from one parent and carry across the gene from the other parent to determine the genotypes and their proportions in the next generation. When we want to extend this Punnett square analysis to a whole population, we must include the genes from *all* parents in this population. When we include all, rather than individual, parents, the proportions of different genes present in the population are represented.

Let's take a hypothetical example of a field of flowers. Let's say we planted a field of pink four-o'clock flowers. Recall from chapter 2 that four-o'clock flower color is a case of incomplete dominance in which heterozygotes **Rr** are pink, homozygotes **RR** are red, and homozygotes **rr** are white. So, if the whole field has pink flowers we know that all the plants have the genotype **Rr**. If these flowers were to pollinate each other, we can predict what the proportions of resulting offspring will be. Remember that **Rr** parents will all generate equal numbers of **R**- and **r**-carrying gametes. This can be represented in a Punnett square as 50 percent, or 0.5 **R** and 0.5 **r**, for all parents (figure 10.2). Then, we multiply the numbers for each of the genotypes to get 0.25, or 25 percent, for each of the four categories of offspring. The interpretation of this Punnett square is that in the next generation in that field, there will be approximately 25 percent **RR**, or red, 50 percent **Rr**, or pink, and 25 percent **rr**, or white flowers. This result would be expected if each flower had equal chances of pollinating and being pollinated, each resulting fertilized flower set its seeds, and all the seeds grew into new plants.

Suppose now we come back the following year to look at this field of flowers that, now, has red, pink, and white flowers in a 1:2:1 (25 percent, 50 percent, and 25 percent) ratio. What frequency of red, pink, and white flowers do we expect to see next year? First of all, what is the frequency, or the proportion, of **R** and **r** genes in the flowers this year? We can calculate it since we know the proportion of the three genotypes:

For 0.25 **RR**   0.25 **R**
For 0.5 **Rr**   since half of 0.5 is 0.25, we get 0.25 **R** and 0.25 **r**
For 0.25 **rr**   0.25 **r**

So, in total, there are 0.5 **R** genes and 0.5 **r** genes, which is what we started with when we had all pink flowers in our field! Let's now cal-

A

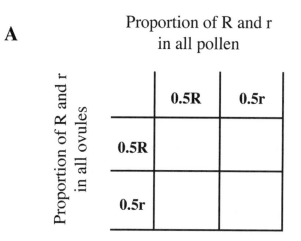

Proportion of R and r in all pollen

B

	0.5R	0.5r
0.5R	0.25 RR	0.25 Rr
0.5r	0.25 Rr	0.25 rr

Figure 10.2 Using a Punnett Square to Calculate Genotype Frequencies. A. The gametes from all male parents (pollen) are shown at the top and gametes from all female parents (ovules) are shown at the side with 0.5 gene frequency. B. We multiply the gene frequencies to calculate the genotype frequencies. This example shows that all four genotypes occur at 0.25 frequency, or 25 percent each.

culate what proportion of different colored flowers we expect the following year. As before, we need to calculate the percentage of **R** and **r** genes in the population. In this case, we observe 25 percent **RR** (or 0.25 **R**), 50 percent **Rr** (0.25 **R** + 0.25 **r**) and 25 percent **rr** (or 0.25 **r**). The ratio of **R** versus **r** is again 0.5 to 0.5, the same as what we found in the previous generation! So, in future generations we expect the same result, that starting with a field of 25 percent red, 50 percent pink and 25 percent white flowers, we again will get 25 percent red, 50 percent pink and 25 percent white flowers. The fact that the proportion of genes and genotypes did not change from one generation to the next demonstrates that gene frequency, or the proportion of different forms of the gene, and genotype frequency both stay the same. This is what the Hardy-Weinberg law states: in populations, gene and genotype frequencies do not change over successive generations. We will see next, however, that certain conditions must be roughly true for this law to be applicable.

Conditions for Observing Constant Gene and Genotype Frequencies

So, what are the conditions important in keeping the gene and genotype frequencies constant? One important assumption is that we expect all the flowers to contribute equally to the next generation. For example, if insect pollinators favored the red over the pink and did not visit white flowers at all, the r form of the gene would be underrepresented in the next generation. This is one type of "selection," and such selection would alter the proportion of the genes that are passed down to the next generation. Also, there should not exist a situation in which different insects only visit one color of flowers. For example, let us imagine that one type of insects only visited red flowers, another type only pink, and another type visited just the white flowers. This is an example of nonrandom mating. In this case, the red flowers would produce only more red flowers, white flowers would only give rise to white flowers, and the pink flowers would give rise to red, white, and pink. Thus, we would get fewer pink flowers than what the Hardy-Weinberg law would predict.

It is also necessary that the wind or the insect pollinators do not bring in pollen from other fields. For example, if the adjoining farm grew all red flowers, because they fetched a better price, and pollen from that field drifted into ours, more of the R gene than predicted would be represented in the next generation. This effect is called "migration," that is, migration of genes from someplace else. The numbers predicted by the Hardy-Weinberg law would also be distorted if pollen from our field were blown somewhere else, that is, migrated away, and became unable to pollinate the flowers in our field. We also hope, and reasonably so, that the mutation rate would be too low to change the frequency of the two forms of the gene. Finally, for the prediction to hold, we must have a reasonable number of flowers in the field. That is, if we had a tiny field with only a few flowers, random chance could alter the proportion of genes that gets passed down to the next generation.

Can we use these predictive tools with humans? Are the conditions of no selection, no migration, random mating, no mutation, and large population size applicable to humans? In fact, for most traits of interest, these conditions do not strictly hold true. Indeed, in the case of genetic diseases, there is clearly selection against individuals with the disease trait. For example, males very sick with hemophilia may not survive long enough to have children. Also, certainly, we would like to think that humans do not mate randomly! And we know further, especially in the United States, that people move quite easily from place to place and even across huge oceans and continents. So, does this mean that the Hardy-Weinberg law demonstrated with the four-o'clock flower is not applicable to humans? No, in many cases, it actually is applicable. This law can be thought of as a "back of the envelope" calculation that gives us an approximate proportion of genes and genotypes in a human population. As long as we keep this in mind, the Hardy-Weinberg law can be a very useful tool.

Another Application of the Hardy-Weinberg Law

Now that we have learned how to use the Punnett square to estimate genotype frequencies in successive generations at the level of a whole population, let us study a general example. Rather than considering 50 percent each of the two forms of the gene as in the four-o'clock plant example, let us try a calculation with 10 percent, or 0.1, of the **a** form and 90 percent, or 0.9, of the **A** form (figure 10.3). Multiplying through, we expect among the offspring 0.01, or 1 percent, **aa**; 0.09 + 0.09 = 0.18, or 18 percent **Aa**; and 81 percent **AA**. If all the conditions mentioned above hold, we expect the proportions of the genes and genotypes to remain the same in the following generations. Let us see if this is true. What are the proportions of **a** and **A** in the second generation with 1 percent **aa**, 18 percent **Aa**, and 81 percent **AA**?

1 percent **aa** gives	0.01 **a**
18 percent **Aa** gives half of 0.18 each for **A** and **a**, or	0.09 **a** and 0.09 **A**
81 percent **AA** gives	0.81 **A**

This gives **a** total of 0.01 **a** + 0.09 **a** = 0.10 **a**, and 0.09 **A** + 0.81 **A** = 0.90 **A**.

A

	0.9 A	0.1 a
0.9 A		
0.1 a		

B

	0.9 A	0.1 a
0.9 A	0.81 AA	0.09 Aa
0.1 a	0.09 Aa	0.01 aa

Figure 10.3 Using the Punnett Square to Calculate the Proportion of Genotypes in the Next Generation Beginning with Gene Frequencies of 10 Percent a and 90 Percent A. A. The gene frequency filled in for all reproductive males and females. B. The whole Punnett square filled in with genotype frequencies.

So indeed, the gene frequency remains constant if we allow all the genotypes to contribute equally to the next generation and do not remove or add genes from another population. And if gene frequencies are constant, we expect the genotype frequencies to stay constant, provided the conditions of no selection, random mating, no migration, no mutation and large population size are generally true. But surely, we did not learn the Hardy-Weinberg law just to talk about hypothetical example of field of flowers! You'll see next that this law is also useful for predicting genotype frequencies for human genetic diseases.

Predicting Gene Frequency for a Recessive Trait

Now, let's see if we can use the Hardy-Weinberg law to illustrate the case of a human recessive disease. As mentioned already, we do expect selection against disease traits, but we can still use the Hardy-Weinberg law to make some predictions. But first, in humans, unlike the examples with flowers, we often do not know the frequency of a recessive gene in the population. The disease trait is phenotypically expressed for a recessive trait only when an individual is homozygous recessive. Thus, the information we can gather is the number of affected individuals in the population, that is, the proportion of homozygous recessive individuals in the population under study. How can we use this information?

Let's again apply the Punnett square (figure 10.4), using as an example the human recessive disease PKU that we learned about in chapter 3. We can fill in the genes and genotypes (figure 10.4.A). Now, we know from medical records that roughly 1 in 10,000 U.S. Caucasians suffer from PKU. This is 0.0001 in decimals, or in scientific notation, 10^{-4}. This number represents the proportion of homozygous recessive individuals so we'll associate that number with the genotype **aa** (figure 10.4.B). The proportion of the **a** gene in the population is thus the square root of the proportion of **aa**, since the number of a multiplied by itself gives **aa**. The square root of 0.0001, or 10^{-4}, is 10^{-2}, or 0.01, or 1 percent. The frequency of the normal **A** gene is thus 1 − 0.01 = 0.99, or 100 percent − 1 percent = 99 percent. We now have the full information necessary to fill out this Punnett square for estimating the frequency of the PKU gene in the population (figure 10.4.D). Therefore, from only knowing approximately how many individuals suffer from PKU, we can estimate the frequency of the PKU gene in the U.S. Caucasian population. This frequency is 1 percent.

An even more useful piece of information we can estimate is the proportion of carrier individuals in the population. As you see from the Punnett square (figure 10.4.D), this proportion is twice 10^{-2}, or 0.02, or, to put it another way, 2 out of 100 individuals (2 percent) are carriers. This may seem to be a surprisingly large number for a disease with a frequency of only 1 out of 10,000 people, but it is correct. This is one very useful aspect of the Hardy-Weinberg law of population genetics: it allows us to calculate the carrier frequency, something that cannot easily be determined in any other way.

A

	A	a
A	AA	Aa
a	Aa	aa

B

	A	a
A	AA	Aa
a	Aa	0.0001 aa

C

	1 A	0.01 a
1 A	AA	Aa
0.01 a	Aa	0.0001 aa

D

	1 A	0.01 a
1 A	AA	0.01 Aa
0.01 a	0.01 Aa	0.0001 aa

Figure 10.4 A Punnett Square Used to Calculate the Frequency of Recessive Disease Genes in a Population. A. Punnett square with just the genes showing. B. As A, with the frequency of **aa** individuals shown. C. As B, with the calculated frequency of the **a** gene calculated. D. Complete Punnett square. To simplify, we write "1 A" rather than ".99 A."

The PKU example also illustrates why it is valid to use this law even when the trait is clearly selected against. (PKU individuals, until newborn PKU testing became widespread, were mentally retarded and unlikely to reproduce). Let us calculate what percentage of the PKU gene in the population is carried by affected homozygous individuals, as compared to the percentage carried by phenotypically normal heterozygous individuals. We know at the outset that there is a proportion of 1 in 10,000, or 0.0001, homozygous recessive PKU individuals, and we calculated the percentage of carrier individuals as 0.02. So the proportion of PKU genes carried by homozygous individuals who have 2 copies of the PKU gene is $2 \times 0.0001 = 0.0002$, and that carried by heterozygous individuals is 0.02. The total is 0.0202. Thus the proportion of the gene carried by homozygous recessive individuals is 0.0002 divided by 0.0202, or ~1 percent. This means that PKU individuals carry less than 1 percent of the total PKU genes in the population, while greater than 99 percent of the PKU gene in the population is carried by phenotypically normal heterozygous individuals. So, even if none of the PKU individuals passed their PKU genes to the next generation, 99 percent of the PKU genes will still be passed onto the next generation by heterozygous carrier parents.

Gene Frequencies Vary in Different Populations

You may have noticed that we stated the frequency of PKU specifically for U.S. Caucasians. Why did we do this? Is the frequency different among different ethnic groups? Depending upon the evolutionary and genetic history of different ethnic groups, frequencies of different genes are different. For example, in the case of PKU, as we saw, the frequency among U.S. Caucasians is approximately 1 in 10,000. However, the frequency of PKU among African Americans is lower, only around 1 in 50,000. Among ethnic Japanese, it is even lower and stands at approximately 1 in 110,000. On the other hand, sickle-cell anemia is much more prevalent among African Americans, at 1 in 400, while the prevalence among the general population is around 1 in 2,500. We will discuss the causes for these differences in the next chapter.

One extreme example of differences in the frequency of a disease among populations is Tay-Sachs disease. This is a non-sex-linked recessive disease that affects the nervous system. It has no cure or treatment, and affected babies die by the age of two or three. Because it is a recessive disease, two phenotypically normal parents, if they are heterozygous for the trait, have a 25 percent chance of having an affected baby (recall chapters 2 and 3). Fortunately, this disease is quite rare in the general population, with a frequency of about 1 in 360,000 or about 3×10^{-6}. We can use the Punnett square to estimate the frequency of carrier individuals in the general population (figure 10.5.A). The square root of 3×10^{-6} is approximately 0.0017, and thus the carrier frequency is approximately 2×0.0017 or about 0.0035, or in other words a little more than 3 out of 1,000 people in the general population are carriers. Thus, the chances are quite small that two carrier individuals will marry and have children. However, it turns out that the frequency of this trait is unusually high among Ashkenazi Jews of Eastern European origin. In that population, Tay-Sachs is found in roughly 1 out of 4,800 individuals or approximately 2×10^{-4}. Also, the values for the general population hold true for the population of New York City, and the values for Ashkenazi Jews of Eastern European origin hold true for members of this community who live in New York City. Clearly, there is not random mating among these groups in NYC, and these groups are reproductively separate. We again use the Punnett square to calculate the frequency of carrier individuals (Figure 10.5.B). The square root of 2×10^{-4} is approximately 0.014, the gene frequency, and the carrier frequency is twice that, or 0.028, or roughly 1 in a 30, or approximately 3 percent, are carriers. This is a much higher carrier frequency than for the general population. The higher frequency predicts a greatly increased chance that two heterozygous individuals from this subgroup in New York City will marry and have children. Tay-Sachs is also unusually high among French Canadians and the Cajuns of Louisiana. For this reason, testing is now available for individuals from these groups to determine if they are carriers.

Newborn Testing and Conditional Probability

How is the knowledge about the frequency of different genes and genotypes in the population useful? As mentioned already, all fifty states test for PKU among their newborns. The accuracy of most of the newborn testing is better than 99.9 percent. Thus, these tests are very accurate but not perfect. Because the tests are accurate, we expect that any baby born with the disease will be detected. However, there is a 0.1 percent chance that even if a baby does not have the disease it will test positive. That is, there is a 0.1 percent chance of false positive. So, let's see the consequences of this false positive result.

Figure 10.6 shows how many babies will test truly positive compared to the false positives. First, we begin by choosing the total number of individuals tested. In Figure 10.6.A we chose 10,000, since the frequency of PKU among U.S. Caucasians is 1 in 10,000. Then we make a row for those that test positive and a row for those that test negative. After that, we make a column for the number of individuals we expect to be affected and for those that are not affected. Only one affected baby is expected, and that baby will test positive. Under the designation "normal," we use the information we have regarding false

A

	1 T	0.0017 t
1 T	1 TT	0.0017 Tt
0.0017 t	0.0017 Tt	3×10^{-6} tt

B

	1 T	0.014 t
1 T	1 TT	0.014 Tt
0.014 t	0.014 Tt	2×10^{-4} tt

Figure 10.5 Using the Punnett Square to Calculate the Proportion of Tay-Sachs Carriers. A. Among the general population. B. Among Ashkenazi Jews of Eastern European origin.

positives to fill in the "test positive" category. The false positive rate is 0.1 percent; 0.1 percent of 10,000 is 10. Thus ten normal babies will test positive! The remainder of the babies will test negative. Now we can calculate the total for those that test positive. Of the 11 that test positive, only 1 is expected to actually have PKU. Any baby that tests positive for any disease is automatically tested again because of these unavoidable false positives. It is highly unlikely that testing the same baby will yield a false positive twice in a row. It is only after a baby tests positive twice that doctors are contacted to check the diagnosis and begin treatment.

Now, let's do the same analysis, using this time the PKU rate among ethnic Japanese. In this population, PKU is found at the much lower rate of 1 in 110,000. We fill in the same table as before in figure 10.6.B, but we choose 110,000 as the total population, since now there is only 1 case of PKU in 110,000 births. The number of affected babies

A

	affected	normal	total
test positive	1	10	11
test negative	0	9,989	9,989
total	1	9,999	10,000

B

	affected	normal	total
test positive	1	110	111
test negative	0	109,889	109889
total	1	109,999	110,000

Figure 10.6 Calculating the Number of False Positives for PKU. A. In the case of the U.S. Caucasian population. B. In the case of the Japanese population.

is just one, and that one will be detected as positive. However, although the accuracy of the test does not change, the false positive rate of 0.1 percent now applies to a much larger population, so the actual number of false positives is 110. So now out of a 111 that test positive, only 1 baby actually has PKU! This conclusion strongly reinforces the necessity of conducting more than a single test to determine which babies are truly PKU positive.

The actual number of disease cases relative to the number of false positives is an important consideration when deciding whether to screen the general population. Another example of this is maple syrup (or fenugreek) urine disease. This is a metabolic genetic disease that can be easily treated with vitamins and dietary control. This disease is quite rare in the general population, and the estimated frequency is about 1 in 300,000. In several years of testing newborns in Iowa, all babies that tested positive turned out to be false positives. Therefore, testing for this disease was discontinued in Iowa in 1995. There are populations that have unusually high rates of this disease, for whom testing is advised. For example, a Mennonite community in eastern Pennsylvania has a rate of 1 in less than 200! We discuss possible reasons for such differences in disease frequency in the next chapter.

A

	X^H	X^h	Y
X^H	$X^H X^H$	$X^H X^h$	$X^H Y$
X^h	$X^h X^H$	$X^h X^h$	$10^{-4} X^h Y$

B

	$1\,X^H$	$10^{-4}\,X^h$	Y
$1\,X^H$	$1\,X^H X^H$	$10^{-4}\,X^H X^h$	$1\,X^H Y$
$10^{-4}\,X^h$	$10^{-4}\,X^h X^H$	$10^{-8}\,X^h X^h$	$10^{-4}\,X^h Y$

1 in 100,000,000 girls are hemophiliac

2 in 10,000 girls are carriers

1 in 10,000 boys are hemophiliac

Figure 10.7 Punnett Squares Used to Calculate the Frequency of a Sex-Linked Trait in a Population. A. Punnett square representing the males at the top and the females on the side. The males are represented by two X chromosomes for the frequency of the normal and disease gene on the X chromosome and by their Y chromosome. The 1 in 10,000 frequency for hemophilia A among males of is shown for $X^h Y$. B. Completed Punnett square with the hemophilia gene frequencies.

Predicting Genotype Frequency for Sex-Linked Traits

We saw in chapters 2 and 3 that we can use the Punnett square to predict the proportion of genotypes in offspring for sex-linked traits (figure 10.7). Similarly, we can use a modified Punnett square to calculate the genotype frequency for sex-linked traits for a population. When we deal with sex-linked traits, the Punnett square has the sex chromosomes represented for the mother by X and X, and for the father by X and Y. The trait we will be discussing is hemophilia, an X-linked trait. Thus, there exists in the population a fraction of X chromosomes with the normal gene, X^H, and another fraction with the hemophilia gene, X^h. Although each individual male in the population has only one X chromosome, to represent all the males in the population, we need to represent males with both X^H and X^h. In order to represent the frequency of the two different forms of the gene on the X chromosome, males are represented by two X chromosomes and a Y chromosome. Now we can insert the information that we have, that approximately 1 in 10,000 males, or 0.0001, or 10^{-4}, in the population has hemophilia A (figure 10.7.A). Since this number is the result of multiplying the frequency of Y (which is 1, or 100, percent among males) with the frequency of X^h, the proportion of hemophilia among males is the same as the proportion of the hemophilia gene on the X chromosomes in the whole population. So, we can indicate 0.0001 or 10^{-4} as both the frequencies of X^h among males (top) and among females (side). Again, because this is such a small number, for ease of calculation we use 1 for the frequency of X^H instead of $(1 - 10^{-4})$. Now we can finish calculating the genotype frequency among the female offspring. Since the genotype frequency in females is calculated by multiplying the frequency of X^h by itself, the frequency of hemophiliac girls is estimated to be about 0.00000001, or 10^{-8}, or 1 in 100 million, and the frequency of carrier girls is about 2 in 10,000. So, just as we saw in chapter 3, we expect and observe that men have a much higher frequency of X-linked diseases than women.

Summary

In this chapter, we extend the rules of Mendelian genetics from individuals to whole populations. We can predict the gene and genotype frequency in a population using a modified Punnett square. The rules of Mendelian genetics extend to populations of people and allow us to predict genotype frequencies by using the Hardy-Weinberg law. The Hardy-Weinberg law shows that in a large population, if all the genes have equal probability of being passed down to the next generation, gene and genotype frequencies remain constant. Understanding how genes are passed down in populations allows us to estimate the proportion of carriers of recessive genetic diseases. There exist cases of very different frequencies of genetic diseases among subgroups of the population. The existence of different gene frequencies suggests non-random mating among these groups. Knowing the predicted frequency of diseases allows us to decide whether there should be population screening for genetic diseases.

CHAPTER 11

Survival of the Fittest?

IN CHAPTER 10, WE SAW that gene and genotype frequencies can vary considerably between different ethnic groups. In this chapter we will explore the underlying reasons for those differences. We will also see that our human activities can effect changes in the gene frequencies of other organisms, from bacteria to insects and other animals. More importantly, many of those changes in turn affect us.

What Is Meant by Fitness?

The phrase "survival of the fittest" may conjure up an image of a big strong lion with his pride of lionesses producing many lion cubs. In that example, the genes of the strong lion, as well as those of the lionesses he chooses, will have a greater chance of being passed down to future generations of lions than the genes of a weak lion without any lionesses. We say that the genes of the strong lion are "selected" because of his ability to father more lion cubs. This, however, is not the only type of selection that exists in nature. Indeed, in the example of a field of four -o'clock plants in chapter 10, we saw a different type of selection. In that case, the preference of insect pollinators could increase the chances of one color gene being favored over another. Thus, selection is a process in which the individuals considered the "fittest" are those that have a higher chance of passing on their genes to the next generation. There are many ways to achieve this. To put it differently, the term "fitness" in this genetic sense does not necessarily refer to strong muscles or the ability to jog several miles. The fittest individuals are not necessarily the strongest, the smartest, or the healthiest. For example, it could be said that white four-o'clock flowers are less fit, not because they are sickly or smaller or have fewer flowers, but because they are not chosen by insect pollinators that prefer red flowers. In such a case, if insect pollinators prefer red flowers, they will not pollinate white or pink flowers, with the result that the white version of the gene will be underrepresented in subsequent generations. Therefore, the genes from the fittest individuals are those that are *selected for* in such a way that their proportion will increase in successive generations. The genes from less fit individuals are *selected against* and their proportion will decrease in successive generations.

Selection Requires Variation

Selection can either be natural or artificial. Natural selection takes place in a natural environment, and selective forces there can be any number of factors such as drought, high or low temperatures, existence of predators, insect pollinators, and so on. Individuals genetically well-adapted to these conditions, or well-adapted to changing conditions, can survive and proliferate. Genetically less-adapted individuals will have a harder time and may well become extinct. How does selection by natural or artificial means take place? Possibilities for selection of some individuals exist because not all individuals in a population are genetically identical and equally fit. This is because each individual, as we saw, is the product of gene shuffling during meiosis, and therefore different individuals do not possess the same gene combinations.

Selective (artificial) breeding for desired characteristics, along with natural selection imposed by the environment, have produced our domesticated animals, pets, and crops. Successful selective breeding also requires variations within populations. That is, if all the animals or plants had the same characteristics, ones that produced more milk or laid more eggs or had higher oil content—or whatever qualities we wanted to enhance—would not exist. *Thus there can be no selection without variation in populations of animals and plants.* The ultimate source of this variation is mutations. Let us now consider a few examples of artificial selection.

Humans have produced many successful strains of domesticated animals and crops. These animals and plant have the highest quality of the desirable properties humans look for. For example, if one observes the milk goats at a local county fair, one sees that they all have tremendously large udders, so large that they would have difficulty walking if let out into the mountainous terrain of their ancestors. These goats were bred for the purpose of high milk production and originated from goats that produced more milk than average. Over time, breeders selected progenies from these goats with even higher milk production and bred them again, until the milk production was much higher than in the original goats, and, by consequence, large udder size was obtained. This example shows that artificial selection by breeders can go against what would have been selected for in a natural environment. In nature, goats with inordinately big udders would surely be selected against, as they would not have the agility of small-udder goats and would be at a disadvantage. Genes responsible for large udders would then quickly be diluted in a population of wild goats.

Selection Can Result in Reduced Genetic Diversity

Thanks to the "Green revolution," the yield of our grain crops, such as rice and wheat, has increased tremendously. Here also, selective breeding of individuals showing desirable characteristics was involved. In some cases, mutants obtained through the use of mutagens were used as progenitors of the high-yield varieties in current production. Artificial selection is generally beneficial to human consumers, but it can sometimes have unforeseen, deleterious consequences.

For example, in our efforts to produce animals and plants with just the desired characteristics, our domesticated animals and crops now have limited genetic diversity. This is due to the fact that, by selectively breeding a limited number of individuals, we stop propagating those individuals that do not possess the characteristics we desire. The rejected individuals and the genes they carry are then at risk for extinction. However, the individuals we select against may possess characteristics of which we are unaware. For example, if new diseases or pests of crops arise, it may well be that the rejected individuals naturally possess genes making them resistant to these pests and diseases, while the selected individuals do not. In other words, through selective breeding, we may have eliminated from the gene pool traits that *may* be important at some point in time under different conditions. This lack of genetic diversity can make the *entire* crop susceptible, and the disease or pest may quickly devastate it.

This is what happened with the Irish potato famine of the 1840s, in which fungi attacked the potatoes. All the potato strains in Ireland were similar and did not have any natural defense again the fungi, and the potato crop was wiped out for several years. In the 1970s in the United States, the corn crop was similarly destroyed because almost all the U.S. corn was susceptible to southern corn leaf blight, another fungal disease.

Similar problems associated with limited genetic diversity can arise in the breeding of pets. Professional breeders are careful to avoid breeding closely related dogs in order to maintain genetic diversity. Nevertheless, for popular breeds of dogs in which there are pressures to produce many puppies, medical problems such as hip dysplasia and behavioral problems have crept into certain breeds. Thus, animals and plants with highly desirable characteristics and limited diversity of other genes may encounter unexpected difficulties. Let us now study a case of natural selection in human populations.

Natural Selection Determined Skin Color in Humans

Selection can occur on Mendelian, or single-gene, traits, but also on traits determined by many genes, since selection acts on the phenotype, as we have seen in the case of goats, lions, and four-o'clock plants. An example of a trait determined by many genes is skin color in humans. Natural selection for or against skin pigmentation is well understood and took place tens of thousands of years ago as modern humans left Africa to colonize the rest of the planet. People from tropical areas with high exposure to the sun are darker in order to protect themselves from the dangers of ultraviolet rays. But ultraviolet rays also serve a necessary function by helping chemicals present in the skin to convert to vitamin D. People with dark skin still make enough vitamin D because high sunlight intensity in the areas in which they live compensates for lower efficiency due to skin pigmentation. On the contrary, individuals from higher latitudes and areas with greater cloud cover have extremely light skin. These individuals use what little ultraviolet light they do get for the purpose of producing vitamin D. You now understand why human variants with light skin pigmentation were not fit to survive and proliferate in the tropics: the harmful effects of ultraviolet light selected against them. In a similar fashion, humans with dark skin pigmentation were at a disadvantage in higher latitudes because they could not make enough vitamin D to stay healthy.

Today, unlike in ancient times, people travel easily throughout the world and carry in their genome the genetic history of their ancestors. Our genes do not change just because we are now in a different geographic area or environment. Light-skinned individuals can now survive in tropical climes because they wear protective clothing and use sun screen. Dark-skinned individuals can now survive in northern climes because they obtain vitamin D from food. Thus, our genetic heritage is a reflection of all the selective pressures that our ancestors have encountered.

Fitness Depends Upon the Environment

Let us imagine a mutation in an insect that made it lose its wings. How would this affect its fitness? We might expect that this would be bad, that is, an insect without wings would have lower fitness compared to the same insect with wings. But what if loss of wings happened to insects that were blown off shore, onto an island, in an area that is often windy? The island is small and the area windy, so it might be safer for insects to walk around rather than try to fly and get blown away into the ocean. Thus whether a particular phenotype is higher or lower in fitness *depends upon the environment or circumstances in which an individual lives.*

The case of insect pests demonstrates the way in which human activity changes the environment to affect selection. Various weevils thrive on agricultural crops and destroy them. Populations of weevils have all types of natural genetic variations because they breed freely in the wild and accumulate random, spontaneous mutations. But farmers started using pesticides to get rid of these noxious pests. When a field is sprayed with pesticides, weevils with natural resistance have higher fitness than weevils without natural resistance to the pesticide. Thus most of the susceptible weevils are killed, and a few weevils, those with natural resistance to the pesticide, survive. What happens after the effects of the pesticides have worn off? Susceptible weevils from surrounding areas that were not sprayed enter the field since there is plenty of food now and fewer weevils to compete with for that food. The net effect of the susceptible weevils "migrating" in from the surrounding fields is to swamp the gene pool with genes susceptible to pesticides. Thus, resistant weevils are again in a minority.

But now consider a field of genetically modified crops that make Bt toxin all the time in the plant tissues themselves (see chapter 6). Unlike the situation above, there is no way for the weevils to get away from the pesticide because it is always present in the crops. Even if more weevils migrate in from the surrounding areas, those susceptible are killed continuously, and, as a result, only resistant weevils survive. These survivors mate and have more weevils. Soon, a higher percentage of weevils have pesticide-resistance genes. There are insufficient numbers of weevils with susceptibility genes surviving to dilute those with resistance genes. Among this large number of resistant insects, some, by random chance, develop even higher resistance to Bt toxin. Of course, those with higher resistance will continue to increase in number over those with less resistance as long as the Bt toxin is present. In the end, only resistant weevils are left, and the pesticide has lost its effectiveness.

This is the scenario that quickly started to unfold with genetically modified Bt crops. Indeed, very soon after the commercial release of Bt-modified crops, farmers began to notice that there were a lot of insect pests on their Bt crops that were not supposed to be there. These insects had become resistant to the Bt toxin. In order to prevent a massive proliferation of Bt-toxin-resistant insect pests, farmers that plant Bt-modified crops must now agree to plant a percentage of their acreage with non-Bt-modified crops close to the field of Bt-modified crops. The acres of non-Bt-modified crops provide refuge for insect pests that are susceptible to Bt-toxin. Without this refuge, the genes for Bt-toxin susceptibility will be eliminated from the population, and Bt toxin, whether sprayed on the fields or made by the crops themselves, will no longer be effective.

Selection and Antibiotic-Resistant Bacteria

All living things are faced with selective forces. In the case of bacteria, human use of antibiotics is a major selective force. We presently have a whole arsenal of antibiotics to fight bacterial infections. But until the 1940s there were no antibiotics at all! Antiseptics such as alcohol and iodine were used well before antibiotics were discovered, but they were not effective against diseases like tuberculosis, pneumonia, and other internal or deep bacterial infections. In the pre-antibiotic days many women chose not to deliver their babies in hospitals because many babies and women died of bacterial infection. Hospitals were particularly unsanitary because doctors easily transmitted bacterial infections among their patients, in spite of all the precautions taken. In World War I, for example, more casualties were due to infections in wounds than to the wounds themselves because there were no antibiotics.

The discovery and subsequent production of penicillin in the 1930s and 1940s dramatically changed the prospect for patients with bacterial infections. Antibiotics became the magic bullet by which bacterial infections could be stopped in their tracks. The 1945 Nobel Prize for Medicine was awarded to Sir Alexander Flemming, Sir Howard Walter Florey, and Ernst Boris Chain for their life-saving work, the discovery and industrial production of the very first antibiotic: penicillin.

So, what has this got to do with selection? Antibiotics provide a tremendous selective force against bacteria, because the function of antibiotics is to kill bacteria. Only those bacteria that can somehow survive an onslaught of antibiotics pass on their genes to their progeny. This can only happen if some bacteria become resistant to an antibiotic. How is this possible? Bacteria have a variety of ways to thwart the action of antibiotics. First, an antibiotic cannot work if it cannot remain inside the bacterial cell. Some bacteria can actively pump out antibiotics and thus survive. Other bacteria can chemically break down the antibiotic so that it is no longer active. Finally, the target of the antibiotic, which is often a bacterial protein, can be modified (mutated) so that the antibiotic can no longer recognize it and poison it. With all these different ways—that is, mutations in different genes—to get around the killing effects of antibiotics, it is no surprise that resistant bacteria arose soon after their introduction.

Penicillin, for example, affects the protective cell wall of a group of bacteria. When cell-wall formation is blocked by the action of penicillin the bacteria burst and die. In the 1940s when penicillin was first introduced, 100 percent of staph infections were effectively treated with penicillin. Now, greater than 90 percent of staph bacteria are resistant to penicillin. A newer antibiotic called methicillin was introduced to counter the rise of penicillin resistance. In 1974, only 2 percent of staph infections were methicillin resistant, but, by 1997, 50 percent of the staph infections had become methicillin resistant. In 2000, over 30 percent of the most common bacteria that cause ear infections in children had developed resistance to penicillin. A similar trend has been found with almost every antibiotic produced to date. So there is an escalating arms race between the bacteria and us. We must continue to develop new antibiotics to keep ahead of the resistance that bacteria inevitably develop.

Why has so much antibiotic resistance appeared? To answer this question, we need to think about selection and survival of the fittest. As patients taking antibiotics, we are exerting selective pressure for antibiotic resistance among the bacteria that infect us. If we take a properly high dose of antibiotics for a sufficient amount of time, chances are excellent that the bacteria causing the problem are completely eliminated. Problems begin when we take an insufficient dose or an incomplete series. Under these circumstances, a few bacteria that survive better than the average, thanks to mutations, are not killed because of the lower dose or shorter time involved. They can now grow up and provide a large number of bacteria that can develop even higher resistance through mutation. These bacteria survive in our bodies and can also find their way into our environment through coughing, sneezing, and defecating. Now if we try to fight this bacterial infection with the same antibiotic, it is no longer as effective as before. The resistant bacteria also can infect other human victims. Then, if these individuals use the same antibiotic, *all* of the invading bacteria will be resistant to that antibiotic. Consequently, the antibiotic will no longer be effective for the population.

Overuse or inappropriate use of antibiotics also contributes to the rise of antibiotic resistance in bacteria. For example, using antibiotics for viral infections does nothing for the infection because viruses are not susceptible to antibiotics. Yet this course of action can promote antibiotic resistance of harmless bacteria that normally reside in our bodies. In and of itself, this does not seem particularly dangerous. However, these antibiotic-resistant harmless bacteria can transmit their genes to pathogenic bacteria as we see below, thereby rendering them resistant as well.

Finally, tremendous amounts of antibiotics are used in agriculture. Antibiotics are used to treat bacterial infections and promote growth in the beef and poultry industries. Antibiotics are also used for plant bacterial diseases. Thus, not surprisingly, bacteria resistant to widely used antibiotics have been found on farms. The particular problem with harmless bacteria or bacteria that are plant pathogens is the way that some bacteria have developed antibiotic resistance. Recall that many bacteria contain a minichromosome, or plasmid (chapter 5). These plasmids can be spontaneously transmitted to different bacterial species. Scientists have discovered a particular type of plasmid that they have named "multidrug-resistant plasmid." This plasmid has several different genes that confer resistance to a number of antibiotics. It can be transmitted from a harmless bacterium to a disease-causing one or from a bacterium that only infects fruit crops to one that is harmful to animals and people. This means its resistance to multiple antibiotics can be transferred to disease-causing bacteria. By all these different means, the rise of antibiotic resistance has reached epidemic proportions. If this situation is not dealt with effectively, we may soon be back to the pre-1940s days when we were without an effective weapon against bacterial infections.

Heterozygous Advantage
We saw in chapter 10 that selection has little effect on recessive disease traits because carriers do not exhibit the trait and therefore are not selected against. For example, in the case of PKU, 99 percent of the PKU genes are found in heterozygote carriers who are phenotypically normal. This means that it takes an extremely long time for recessive traits, even lethal ones, to be eliminated from the population. Indeed, as the recessive form of the gene is eliminated in the form of unfit, homozygous affected individuals, a greater percentage of the disease gene is carried by phenotypically normal heterozygous individuals. Was there a reason for such disease traits to appear in fairly high numbers to begin with? Why are there differences between different ethnic groups for these and other diseases? We are finding out that the answers to these two questions reside in the fact that for many recessive diseases there is a heterozygous advantage to carrier individuals.

The classic case is that of a disease we already mentioned in chapter 3, sickle-cell anemia. Recall in the last chapter that the incidence of sickle-cell anemia is much higher (about 1 in 400) among African Americans than among Caucasian Americans, who show a frequency of around 1 in 2,500. Why is there such a disparity between these two ethnic groups? The answer lies in the genetic history of these two populations. African Americans came from tropical regions with high incidence of malaria, whereas most Caucasian Americans came from temperate regions in Europe, where malaria was not prevalent.

Individuals who are heterozygous for the sickle-cell trait do not, of course, exhibit sickle-cell anemia because the trait is recessive. However, these heterozygous individuals, it turns out, are more resistant to malaria. Malaria is a microbial infection spread by mosquitoes that carry the malaria parasite. When mosquitoes carrying parasites bite a person, the parasites enter the blood stream and the red blood cells of this individual. The parasites infecting the red blood cells make the inside of the cells acidic. This does not affect normal β-hemoglobin in the red blood cells of a person homozygous for the normal β-hemoglobin gene. As a result, the red blood cells of this person retain their normal shape, and infection by the parasite can continue. Death results if the infection is left untreated. However, if the red blood cells of an individual heterozygous for the sickle-cell trait becomes acidic, the red blood cell takes on a sickle shape due to the mutated β-hemoglobin. The abnormal shape of the red blood cells signals the body to eliminate them. Thus, the malaria infection cannot proceed and the infected person survives.

This mutation, then, in the heterozygous state, provides a selective advantage to people living in areas where malaria is prevalent. Of course, it does not provide any advantage to people who do not experience malaria. The sickle-cell trait is debilitating to homozygous individuals and is thus a selective disadvantage. The selective advantage it provides to heterozygous individuals offsets this disadvantage and maintains a higher proportion of sickle-cell trait in the population than one would otherwise expect. This is a classic example of heterozygous advantage that maintains a disease gene in a population.

More recently, a mutation in another gene called G6PD (glucose-6-phosphate dehydrogenase), which codes for an enzyme necessary for red blood cells to obtain energy from glucose, has also been found to exhibit heterozygous advantage. Certain mutations in G6PD, which in the homozygous state causes anemia, also provide resistance to malaria infection in the heterozygous state. Similarly, mutations in other genes for proteins found in red blood cells are thought to confer selective advantage in resistance to malaria. Malaria may have been quite prevalent for a long time and thus a strong selective force acting on the genetic makeup of peoples from tropical regions.

A last example of heterozygous advantage is that of PKU, a disease you have heard about already in chapters 3 and 10. As with sickle-cell anemia, homozygous, recessive individuals are highly disadvantaged.

SURVIVAL OF THE FITTEST? 163

However, again as with sickle-cell anemia, scientists postulate that the PKU trait provides a selective advantage to heterozygous individuals. The mild, damp climate of the British Isles is conducive to the growth of molds in grains and other stored foods. These areas also suffered repeated widespread famines. When one is starving, moldy food is better than no food. Molds have toxins that, among other things, cause spontaneous abortion. However, women heterozygous for the PKU trait appear to have had a lower spontaneous-abortion rate than women homozygous for the normal gene. Thus, the presence of the PKU trait in a heterozygous condition could have favored the survival of offspring from heterozygous women who resorted to eating toxic, moldy food. This would have favored the persistence of the PKU gene in populations where a combination of periodic starvation and moldy foods existed.

As we understand more about the evolution of genetic traits in humans, other recessive genetic diseases for which there is heterozygous advantage may be recognized. For example, cystic fibrosis is another disease that is postulated to confer heterozygous advantage in areas with diarrheal diseases.

Why Do Dominant Genetic Diseases Exist?

We have seen that deleterious traits can remain in the population if they are recessive and if they also provide heterozygous advantage. But we also know of dominant genetic diseases. How do these conditions appear and remain in the population? Dominant traits initially appear as do all other genetic variations: by mutation. In fact, some dominant genetic diseases show a high rate of new mutations; that is, they are not inherited from the parents but appear anew. For example, Marfan syndrome is a dominant genetic disease caused by a mutation in a connective-tissue protein called fibrillin. The gene encoding this protein is about 200,000 base pairs long, quite a large gene. Because most of this protein's structure is important for its proper functioning, the large gene that codes for it provides a large target for mutations. It is estimated that approximately 25 percent of Marfan cases are due to new mutations.

Scientists are identifying more dominant mutations as time goes by, but the great majority of these mutations are extremely rare. Some mutations are probably not recognized as dominant because the individual with the dominant mutation does not survive to pass it on. The first dominant mutation found in humans, brachydactyly, was recognized as such because of one large family affected with it. Similarly, other rare genetic diseases are only recognized when a large family with a particular phenotype is recognized and pedigree analysis is performed.

Yet we find that some dominant diseases are more common than one might expect from a rare mutation and a few families passing on this trait. This is the case for Marfan syndrome, Huntington's chorea, and familial early-onset Alzheimer's disease. The three diseases share a common feature: individuals with the disease traits do not express the disease phenotypes until they are past their reproductive age. That is, by the time the phenotype appears, individuals may have already passed on their defective genes to the next generation. Thus, selection does not have a chance to act because affected individuals reproduce before the disease is apparent. Selection can only act on the phenotype, and selection manifests itself in individuals' success or failure in passing on their genes.

Small Populations

We saw already that, depending upon the genetic history of different peoples, the frequency of genetic traits can differ dramatically among different populations. This difference among groups is accentuated by small population size. Two phenomena, the founder effect and random genetic drift, account for the effects of small population size on gene frequency.

The founder effect is so named because it is observed when a small group of individuals founds a new population. This effect can also be observed when there is a sudden reduction in population size. To illustrate this concept, imagine a very rare genetic disease with a frequency of 1 in 100,000 people. Now, let us imagine that from this population, 100 people decide to emigrate to an unsettled part of a territory. Let us further imagine that, just by chance, the one individual with the disease gene is among the hundred individuals who leave the group. So, just by random chance, the frequency of disease in this new population jumps to 1 in a 100 instead of 1 in 100,000! Thus, without selection, since no one died before passing the gene on, all of a sudden just by subdividing into a small population, the rare trait is now not so rare. This is called the founder effect because it is particularly evident when a founding population colonizes a new land. However, the effect is no different if the population was suddenly subdivided or decimated. For example, assuming the same frequency of 1 in a 100,000, if a huge volcano erupted and almost everyone was killed except the same 100 as above, we would again end up with a 1 in 100 frequency.

Box 11.1 *DNA Sequences Provide Clues to Human Evolution: The Founder Effect in Prehistoric Africa*

DNA sequencing has been used extensively in the study of human origins and evolution. By looking at DNA sequences from different organisms and measuring how much these sequences differ, it is possible to determine the time of appearance of a species. The rationale is that variations in DNA sequences are due to mutations, and the longer the period of time that passes, the more mutations that can accumulate. By comparing the DNA sequences of homologous genes obtained from various individuals belonging to different groups, we can measure how different the DNA sequences are and thus how much time has elapsed since the two lineages split. That is, sequence comparisons allow scientists to estimate the time when the two groups shared a common ancestor.

One particularly convenient piece of DNA to sequence for determining evolutionary relatedness among humans is mitochondrial DNA. In addition to nuclear DNA, animal cells contain DNA in bodies called mitochondria. Mitochondria are the "energy factories" of cells and contain a bit of genetic material. Because mitochondrial DNA is relatively short—about 16,600 base pairs in humans—it is easier and faster to sequence than the whole nuclear genome. Also, portions of mitochondrial DNA mutate more rapidly than many of the genes in our nuclei, in a time span of a few thousand years.

Mitochondrial DNA from various human ethnic groups was sequenced, and these sequences were used to reconstruct an evolutionary tree. The tip of the branches represents the ethnic groups living today. The groups that branch out closer to the base of the tree (the root) represent the earliest divergences of the human ethnic groups. It turns out that people from sub-Saharan Africa appear at many different branches of the tree. A group consisting of only sub-Saharan Africans branch first from people from other continents. One interpretation of this result is that

continued on next page

Box 11.1 *continued*

modern human beings, *Homo sapiens sapiens,* first appeared in Africa about 150,000 years ago (Figure B.11.1). These humans subsequently migrated out of Africa and colonized the whole world. There are more different DNA sequences from people in Africa than elsewhere on earth because humans were in Africa first. Interestingly, mitochondrial DNA is inherited only from mothers. An egg contains lots of mitochondria, while the sperm head that fer-

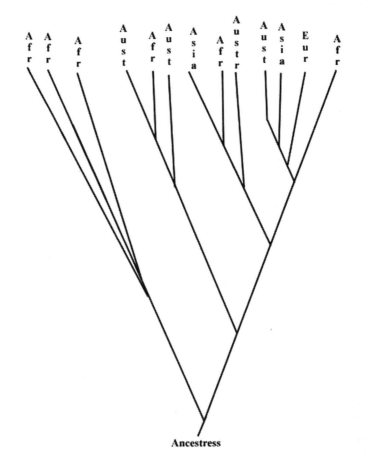

Ancestress

Figure B.11.1 Simplified Evolutionary Tree of Humans. This evolutionary tree of humans was derived from the study of mitochondrial DNA by Allan Wilson. The key features include a group of Africans branching out from near the root of the tree; Africans are also represented on all the other branches.

Box 11.1 *continued*

tilizes the egg does not. This has led to the intriguing concept of "Eve," the first group of African women whose mitochondria is still (with base pair changes) within all of us. "Eve," however, should not be considered a single woman; rather, there probably were thousands of "Eves" that gave rise to modern humans. Nevertheless, mitochondrial DNA sequences fit well with the idea of a founder effect at the root of the human evolutionary tree.

DNA sequence comparisons were also made on the DNA from the Y chromosome. As we know, this chromosome is found only in males and thus transmitted only through fathers. Studies of the differences in sequence on the Y chromosome among different ethnic groups are similar to those obtained with mitochondrial DNA. Here also, scientists have interpreted this evidence as African "Adams" from whom humankind is derived. These "Adams" and "Eves" migrated out of Africa to Europe and Asia and finally to the Americas. The thousands of years during which the first humans traveled to all corners of the world allowed for natural selection and genetic drift to occur. Natural selection worked at the level of skin color and general body build. Random genetic drift and founder effect explain DNA sequence variations that do not seem to have selective advantage or disadvantage. Thus, Y-chromosome studies also support an African founder effect for the human family.

In addition to revealing the origins of *Homo sapiens*, mitochondrial DNA sequencing may shed some light on the fate of an extinct branch of the human family. It is known that the modern humans that colonized Europe and the Middle East (they are often called Cro-Magnons, after the place in France where their fossils were first discovered) encountered other humans, the Neanderthals (*Homo sapiens neanderthalensis*). These Neanderthals were well adapted to the cold climate prevalent during the last Ice Age. However, as the climate warmed up, Cro-Magnons progressively took over the Neanderthals. The latter became extinct about 30,000 years ago. A question that has intrigued paleontologists is how Neanderthals became extinct. Two theories are proposed: one that Cro-Magnons eradicated Neanderthals through warfare, while the other theory holds that Neanderthals and Cro-Magnons interbred, to result in

continued on next page

Box 11.1 *continued*

today's European and Middle Eastern populations. Researchers were able to isolate mitochondrial DNA from a limited number of Neanderthal fossil bones. Sequencing this DNA indicated no close relationship between modern human mitochondrial DNA and Neanderthal mitochondrial DNA. This seems to rule out the interbreeding theory. Therefore, it may well be that our ancestors committed genocide on the Neanderthals to occupy the ecological niches previously held by them.

Random genetic drift is a process that occurs in a small population after the founding of that population. As the name indicates, the process is random, that is, it is *not* due to a fitness advantage or disadvantage. We will demonstrate this with a real-life example of the group represented by Old Order Amish Mennonite Church. This group arose in central Europe during the seventeenth century and began migrating to the United States in the early eighteenth century, predominantly to Pennsylvania and Ohio. Those who espouse the traditional teachings of this church wear plain clothing: men wear hats and do not trim beards, while women wear long dresses, capes, and bonnets. They also use horses and buggies rather than automobiles and generally shun modern conveniences. These people stay in relatively small communities and tend to socialize within their groups. The necessary ingredients for observing random genetic drift is present in this situation, a small founding population that tends to marry within its group. By emigrating in small numbers, the group has set up a founder effect. For example, the frequency of dwarfism, a recessive trait, in the general population is roughly 1 in a million. By using the Hardy-Weinberg law that we learned in chapter 10, we can calculate that in the general population there are roughly two in 1,000 who are carriers (figure 11.1.A). However, *just by random chance*, two carrier heterozygotes were present in a founding population of one hundred Amish emigrants. So, the founder effect increased the frequency of dwarfism carriers from two in a 1,000 to two in 100 in this particular emigrant group. Using the frequency of carriers of two in 100 we can calculate how many

A

	1 A	10^{-3} a
1 A	1 AA	10^{-3} Aa
10^{-3} a	10^{-3} Aa	10^{-6} aa

B

	A	0.01 a
A	AA	0.01 Aa
0.01 a	0.01 Aa	aa

C

	A	0.01 a
A	AA	0.01 Aa
0.01 a	0.01 Aa	10^{-4} aa

D

	A	0.27 a
A	AA	Aa
0.27 a	Aa	0.07 aa

Figure 11.1 Frequency of Genes and Genotypes in a Small Population. A. In the general population, dwarfism is quite rare and found in 1 out of a one million, or 10^{-6}, the frequency of **aa**, since dwarfism is a recessive trait. We can calculate the frequency of **a**, the dwarf form of the gene, as 10^{-3} and the frequency of carriers as 2 in 1,000. B. We calculate the dwarfism gene frequency in an emigrant population of 100 that just by chance has two heterozygotes (**Aa**) among them by putting 1 in a 100 or 0.01 in both the boxes for the heterozygous carriers. We see that 0.01, or 1 percent, is the frequency of the **a** gene in this smaller population. C. We now can calculate the expected frequency of dwarfs (**aa**) in this new population as 0.0001 or 10^{-4}. That is, we expect that with the dwarfism gene's frequency of 1 percent, there will be 1 dwarf in 10,000 births. D. Now we use the actual frequency of dwarfism observed in this population, 1 in 14, which is 0.07. By finding the square root of 0.07, we get 0.27, or 27 percent, as the estimated frequency of the **a** gene in the population. Note that this is significantly higher than the 1 percent frequency of the **a** gene observed in the founding population. This is genetic drift: that is, just by random chance in a small population, the frequency of a deleterious gene can actually increase. Note that this Punnett square is incomplete because the frequencies of **A**, **AA**, and **Aa** are not calculated.

homozygous dwarf individuals we expect to find in this Amish village (figure 11.1.B). This frequency is expected to produce 1 dwarf in 10,000 people. Note that this number is much higher than 1 in 1,000,000, as is found in the general population. However, we actually know that in this particular Old Order Amish community the frequency of dwarfism is 1 in 14. This indicates that the frequency of dwarfism gene in this community is higher today than it was when the population was founded. From this, can you calculate the expected frequency of carriers in the population?

We can use the Hardy-Weinberg law to calculate the frequency of the dwarfism gene in this community, which turns out to be approximately 25 percent (figure 11.1.C). How did this happen? Clearly, dwarfism, especially for a group of people who do not believe in using modern conveniences, is not advantageous. This increase in dwarfism gene was due to *random genetic drift*. Just because the population size was small, *random* effects changed the frequency of genes, even genes that might have a selective disadvantage. Indeed, an extensive study of a number of Amish communities shows an unusually high incidence of different genetic traits, not just dwarfism. That is, an unusually high frequency of genetic traits that are rare in the general population exists in the Amish community, for example, traits such as the neuromuscular disorder called Amish nemaline myopathy, and Amish brittle hair syndrome.

Founder effect and random genetic drift affect all living creatures that experience a dramatic decrease in population size. For example, it is postulated that the massive hunting pressure that reduced the population size of lions may have contributed to the loss of fertility among these animals. It is expected that other endangered species reduced in number by loss of habitat or hunting will face these same random effects of small population size. Thus, even deleterious genes may *increase* in frequency in populations that are already in danger, just based on the random nature of passing on genes in a small population.

Summary

In the context of genetics, fitness means the ability to pass on genes to the next generation. This may be due to the strength and fecundity of some individuals, but it can also be the result of attraction to mate or pollinator. Selection is the force that favors or disfavors some phenotypes relative to others under a certain set of environmental circumstances. For selection to act, there must be variation of traits within a population. Without this genetic diversity a population is ill prepared for changes in its environment. Selection can be natural or artificial. Artificial selective forces such as antibiotics and pesticides can quickly increase resistance among bacteria and insects. Human genes reflect the history of the selective forces experienced by our ancestors. For example, some deleterious recessive genetic traits exist in higher proportion than expected due to the heterozygous advantage they confer. Also, some diseases are more prominent than expected, not because of any selective advantage, but because of the founder effect or random genetic drift. Thus, small populations, just because of their size, can contain a high proportion of less-fit individuals.

Try This at Home: Demonstrations of the Effects of Small Population Size

1. Founder Effect

The founder effect can be demonstrated in a fun, tasty way. Any individual-size packages of candy pieces, such as small bags of M&Ms, will work. The bags of candy must have different colors of candy pieces and have relatively few pieces in each individual bag.

Now, open one package and count how many pieces there are of each color of candy. You may eat all the candy once the numbers have been noted. Then do this with a few more bags of the candy. Avoid the danger of this exercise by having your friends and family members help out, at least in eating the candies! At this point, tally the number of different colors from each bag. Now estimate the ratios of the different colors of M&Ms at the factory by adding the number of each color of M&Ms in *all* the bags together and determining their ratio. Now compare this ratio for the "factory" with the ratio of different colored M&Ms in each individual bag.

This exercise demonstrates what is known as the founder effect in population genetics. Imagine that the M&Ms represent individuals in a population with different colors representing different forms of a gene. In the large population, at the M&M factory, each color is represented by many individuals. Yet when they put just a

continued on next page

few M&M into each of these bags, *just by chance* more of some colors and fewer of others are put in the bag. Each bag has a different random number of colors. Similarly, *just by chance* in a population, a rare genetic disease trait may appear.

2. Random Genetic Drift

This next exercise demonstrates random genetic drift. As with the founder effect, drift is more evident with small populations because random chance has greater impact on small populations than on larger ones.

Begin with a large number of pieces of colored paper, let's say green, purple, and orange. We suggest at least 100 pieces, with 60 percent green, 20 percent purple, 20 percent orange. This is your original "genetic pool," wherein each different form of the gene is represented by a given color and each piece of paper represents an individual.

Then pick ten pieces of paper randomly (that is, blind, without regard to color) from this initial "genetic pool." This is generation 0. Keep track of how many of each color you have at each generation. Note that the chances are that you did not get the same proportion of colors as in the original "genetic pool" of six green, two purple, and two orange. This, again, is the founder effect. That is, *just by chance*, you might pick many more purple or orange pieces than is representative of the original pool, but, just as likely, you might not get any purple or orange.

Now, we will reproduce this population for a few generations. At "each generation," randomly pick a piece of paper from your small population of ten. After you pick, note the color, then return the piece of paper to the population of ten, mix the pieces of paper, and then pick another and note its color. Do this ten times so that you again have ten individuals. This process simulates the genes that are randomly passed onto the next generation. The number of different colors represented by *these* ten individuals may or may not be the same as the original 10 pieces. But now, these ten colors you picked will form generation 1. If the proportion of colors is the same as generation 0, just use the same pool of ten individuals to make a new generation. If the proportion is different, then get more of the colors you need to form a new generation representing the colors you picked. This is generation 1, from which you will again pick ten individuals to form generation 2 and so forth until you have ten generations.

What you have done is simulate ten generations for a very small population of ten individuals. To visualize the result, you can plot the number of pieces of the three different colors on the y-axis of a graph as a function of the generation number on the x-axis, as shown for the two sample simulations in figure B.11.2 and in table 11.1. What happened to the numbers of different colors in your simulation? If this simulation was repeated, what might you expect?

Note in the example shown that sometimes one color can become predominant or even become the only color in the population. That is, if one color is eliminated, there is no way for that color to reappear in the population without mutation (which occurs at a very low rate) or migration. If this happened in a real population, it would mean that genetic diversity would dramatically decrease. A gene trait that is not highly represented in the original population can be eliminated, as in simulation #1 with purple, and even traits that are prevalent in the original population can disappear, as with green in simulation #2. Conversely, a rare trait can be the only one in the population, as with purple in simulation #2.

Table 11.1 Results from Two Simulations of Random Genetic Drift

Simulation #1

generation	#0	#1	#2	#3	#4	#5	#6	#7	#8	#9	#10
green	4	4	4	6	7	7	8	7	8	9	9
purple	1	1	2	1	0	0	0	0	0	0	0
orange	5	5	4	3	3	3	2	3	2	1	1

Simulation #2

generation	#0	#1	#2	#3	#4	#5	#6	#7	#8	#9	#10
green	4	3	4	3	4	2	2	1	1	0	0
purple	5	6	6	7	6	8	8	9	9	10	10
orange	1	1	0	0	0	0	0	0	0	0	0

continued on next page

Simulation #1

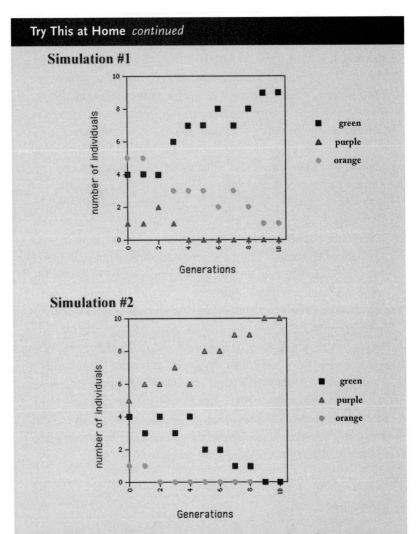

Simulation #2

Figure B.11.2 Simulations of Random Genetic Drift. Each form of the gene is represented by colors of paper represented by different symbols on the graph. The frequency of different forms of the gene at each generation is shown. Though we began with the same proportion of the colors, just by random chance, the proportions of purple and orange are different. This is the founder effect. Then with each generation, because the population size is small, by random chance certain colors can predominate in the population or even be the only color in the population.

CHAPTER 12

Nature Versus Nurture

SO FAR WE HAVE SEEN THAT MENDELIAN TRAITS are categorized into discrete classes. That is, the different phenotypic classes
that result from a cross can be sharply distinguished from one another, whether they are the colors of flowers or disease traits in humans.
We also mentioned that many traits of economic and medical importance do not fall into discrete categories with sharply distinguishable
phenotypes. These include the yields of most crops and farm animals
and many common diseases like heart disease and diabetes. You may
recall that Correns, one of the rediscovers of Mendel's laws, mentioned that even with plants that exhibited some Mendelian traits,
other traits were not explainable by the simple Mendelian rules of inheritance (see chapter 2). We now know that these other traits are *polygenic* traits, or traits due to the contributions of many genes. These
polygenic traits exhibit more complex inheritance patterns. They also
happen to be common and important in the animal and plant king-
doms, including humans. In this chapter we will learn that genes contributing to polygenic traits do follow Mendel's rules in spite of their
complexity.

Polygenic Traits Are Additive

The traits that we will study in this chapter are *not* under the control of single genes, as in the case of simple Mendelian traits. Complex traits have been given many names, including polygenic, multigenic, multifactorial, and quantitative. They are called quantitative, or measurable traits, since they are not either/or types of traits, such as white or red flower color, but show relative amounts of a particular feature. These traits are said to be additive because each of the many genes that contribute to the trait are added together to determine the phenotype. To illustrate this, let's go back to the example of incomplete dominance in a single-gene trait of the four-o'clock flower seen in chapter 2. Recall that in this case, the genotype of the red flower is **RR**, the white flower is **rr**, and the pink flower is heterozygous **Rr**. One way to think of how the color of the flower is determined is to count the number of **R** genes in these individuals. As you can see, the darkness of the flower (red or pink or white) is determined by the number of **R** genes. Indeed, the darkest flower (red) is obtained with an **RR** combination. **Rr** flowers, with only one **R** gene, are pink, and the **rr** flower are the palest, white. Thus, the **R** genes exhibit additivity because the intensity of the phenotype increases with the number of **R**s. Of course, in this case, only two, one, or no **R** genes are possible.

Now let us imagine a situation in which *two* genes, each existing in two possible forms, behave in the same way. Figure 12.1 shows a Punnett square of two heterozygous individuals mating to produce offspring. These individuals are both designated **AaBb**. This example is similar to the Punnett square we drew in chapter 9 when we discussed the behavior of two genes in a cross. However, the difference in this case is that the two genes contribute to the *same* phenotype and not different phenotypes as previously. As before, we see that there are nine different genotypes represented in the sixteen boxes of the Punnett square. But remember that these two genes contribute to the same phenotype, in this case red flower color. For simplicity, suppose that each of the uppercase genes, **A** and **B**, contributes the same amount of pigment to the phenotype. The **a** and **b** genes contribute no pigment. Then you can see that the nine different genotypes represent five different phenotypic classes, from dark red to white, that vary according to the number of uppercase genes present in the organism. This number varies from four (the **AABB** genotype) to zero (the **aabb** genotype).

	A B	A b	a B	a b
A B	A A B B	A A B b	A a B B	A a B b
A b	A A B b	A A b b	A a B b	A a b b
a B	A a B B	A a B b	a a B B	a a B b
a b	A a B b	A a b b	a a B b	a a b b

Genotypic classes	#	Phenotypic classes	#
A A B B	1	4 A or B	1
A A B b	2	3 A or B	4
A a B B	2		
A A b b	1		
A a B b	4	2 A or B	6
a a B B	1		
A a b b	2	1 A or B	4
a a B b	2		
a a b b	1	0 A or B	1

Figure 12.1 Inheritance of Two Pairs of Unlinked Additive Genes. Punnett square showing the inheritance of two forms of two different unlinked additive genes from double heterozygous parents. Note that among the sixteen boxes, there are nine different genotypes represented. If the upper case form of each gene contributes equally to the trait, the nine different genotypes represent five phenotypic classes in the proportions listed.

Since we assumed that both uppercase genes, **A** and **B**, contribute the same amount of red pigment to a flower, it follows that the **AABB** genotype will be darkest red and **aabb** will be white. Intermediate numbers of uppercase genes represent intermediate phenotypes from reddish to pinkish. For example, the **AaBB** genotype will be phenotypically very close to the **AABB** flowers, only a slightly paler red. The same effect will be observed with an **AABb** flower. This is because both genotypes contain three uppercase genes. On the other hand, **Aabb** and **aaBb** will show the same shade of color, light pink. Of course, **aabb** will be white because it does not contain a single uppercase gene. So, unlike the case of the additive single-gene trait that showed three phenotypic color classes (the four-o'clock flowers), we can expect to see five different color classes in the case of two genes contributing equally to flower pigmentation. The larger the number of genes that contribute to the same phenotype, the larger the number of phenotypic classes and the finer the gradation between these classes.

With polygenic traits, we often do not know the number of genes that contribute to the trait, nor the quantity and quality of contribution of each gene to the phenotype. For example, in the case of human skin color, we are not all different shades of black to white. In fact, skin color ranges from very pale white to reddish, brownish, yellowish, and a very dark tinge. This means that each gene that contributes to our skin color may have more than two forms, and each form of the gene may contribute different amounts of different colors of pigmentation. Thus, although each of the underlying genes is being inherited in a Mendelian fashion, we cannot analyze the inheritance of the trait using Punnett squares.

Polygenic Traits Exhibit Continuous Variation in Phenotype
In the case of a single incomplete dominant trait, we observe three phenotypic classes in a 1:2:1 ratio (figure 12.2.A). In the above case of two genes, wherein each gene contributes equally to the trait, the nine different genotypic classes fall into five different phenotypic classes. These phenotypic classes are represented in a 1:4:6:4:1 ratio (figure 12.2.B). A similar situation, with more genes, will result in more phenotypic classes. The ratios for the phenotypic classes are shown in Figure 12.2.C and D. If we continued this process further for more genes, in which each gene provides an equal contribution to the phenotype, we would get a bell-shaped curve, or normal distribution, of many

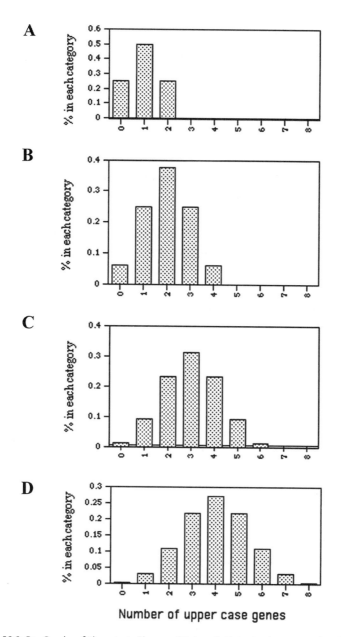

Figure 12.2 Bar Graphs of Phenotypic Classes of Polygenic Traits. In these examples, the up-percase forms of the genes each contribute equally to the trait. Graphs depict the percentage of individuals in each phenotypic category. The categories represent the total number of upper-case genes, from none on the left to eight on the right, for a trait determined by up to four different genes. A. A graph depicting a single incomplete-dominance gene trait. B. A graph depicting the two-gene trait shown in figure 12.1. C. A graph depicting a three-gene trait. D. A graph depicting an example of a four-gene trait

more phenotypic classes. In other words, phenotypes would no longer be sharply definable.

In the examples given above of flower color, each gene contributes equally to the phenotype, but this is probably not the case in real life examples of polygenic traits, where many genes control the phenotype. You can imagine that with a potential for varied genetic contribution, even with a few genes, rather than observing three or five distinct phenotypic classes, we would see an almost continuous variation. *This continuous variation in phenotype is the hallmark of a polygenic trait.* If we graphed the number of people of different heights, we would get a graph similar to a normal distribution: most people would be of average height, and there would be fewer people that are taller or shorter and few that were very tall or very short. Weight or skin color in humans, size of fruit in plants, and oil content of canola seeds are all polygenic traits that exhibit continuous variation in phenotype. *Thus, without knowing anything about the genes underlying a trait, if there is a continuous variation in the trait in the population, we can safely assume that many genes determine the expression of that trait.*

Polygenic Traits Are Influenced by the Environment

You now know that polygenic traits are more complicated than single-gene traits just because many genes are involved. In addition, we often do not know how many genes there are nor the different qualities and quantities contributed by the genes to the expression of a trait. Complicating this situation even further is environmental conditions' ability to influence the expression of these traits. For example, let us take the case of human skin color. The genes that contribute to our skin color also contribute to the tanning qualities of our skin. When exposed to sunlight, a light-colored person may be prone to burn and not tan. Another person may be similarly light colored, but his or her genes may allow better tanning. Such a person thus may appear darker after exposure to sunlight.

A striking example of interaction between genes and the environment is that of the wild plant *Achillea millefolium*. As is the case for many other plants, *Achillea* can be propagated through cuttings. Because cuttings do not involve sexual reproduction, cuttings from a single plant have the same genotype as the original plant; that is, they are clones. Researchers were interested in knowing how this plant grew at different altitudes and whether different genotypes responded differently. Growth is known to be under the control of many genes and influenced by the environment. It was observed that cuttings obtained from one plant grew well at high and low altitudes but did poorly at an intermediate altitude. However, cuttings from another *Achillea* plant, possessing a slightly different genotype, grew almost equally well at all altitudes. This example shows that interactions between genes and the environment can be complex and unpredictable.

Measuring Variance in Traits and Estimating Heritability

The differing degree of genetic versus environmental contribution to a trait is called heritability. Many Mendelian, or single-gene traits are unaffected by the environment. An extreme example is our blood type, which is unaffected by the environment and thus is 100 percent heritable. For polygenic traits, genetic factors contribute a portion of the trait, and the environment contributes the rest. If we wish to select animals or plants for a given trait, heritability is an important consideration. Heritability predicts how successful we will be in selecting for a trait by breeding individuals with the favorable trait. A trait that is highly heritable can be selected for, whereas if a trait is mostly determined by the environment, selection will not help in producing animals or plants with that trait.

Many traits of agricultural crops and animals, such as milk production in cattle and oil content in corn, are polygenic traits. We saw that polygenic traits such as these do not produce easily distinguishable phenotypic categories. So how do we measure the phenotypic differences brought about by polygenic traits among individuals in a population? How do we estimate the influence of the environment on these traits when phenotypes are the only thing we can directly observe?

If we measure traits from many individuals and plot them, polygenic traits approximate normal, bell-shaped distributions, as we saw in figure 12.2. This type of distribution can be analyzed statistically using two measures, the mean and the variance of these plots. The mean (or average) is the sum of all the measurements of a trait in a population of organisms (for example, weight), divided by the number of organisms measured. Variance is the flatness or the sharpness of the normal distribution around the mean. Figure 12.3 illustrates these concepts. You can see in this figure three normal curves that have the same mean but are very different in their spread. The flatness of the curve is indicative of the spread of the measurements, or variance.

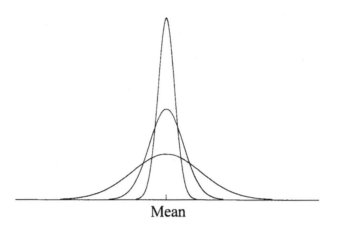

Mean

Figure 12.3 Overlapping Graphs of Normal Distributions or Bell-Shaped Curves. All three distributions have the same mean but very different variances. The sharpest one has the least variance and the flatter ones have higher variance. Plots of measurements of polygenic traits from many individuals would have shapes similar to these. Both genes and the environment contribute to the variance in plots of polygenic traits.

In the case of polygenic traits, the variance observed for a particular phenotype (V_P) depends on the variance due to the genotype (V_G) plus the variance due to the environment (V_E). In order to determine heritability, defined as the genetic contribution to the phenotype, researchers try to make the environmental variance as close as possible to zero, that is, to keep the organisms under study in the same environment. With animals and plants, this can be done by growing these individuals under conditions that are as close as possible for all the individuals under study. Under constant environmental conditions, which mean a small V_E, almost all the variance is due to the genes. Conversely, if one wants to know the contribution of the environment, one can raise in a natural setting highly inbred, thus genetically very similar, animals and plants characterized by small genetic variance, V_G. These estimates of variance due mostly to genetics or environment are then used to determine heritability, that is, the genetic contribution to a polygenic trait.

Determining heritability involves producing many offspring and detailed bookkeeping of the trait in question to allow for statistical analysis, as described above. For example, we may wish to increase egg production by either selecting chickens that produce larger eggs or ones that produce more eggs during the year. It was determined that for white leghorn chickens, heritability for egg weight was 60 percent whereas heritability of number of eggs per year was only 30 percent. This means that the trait for the weight of eggs has a larger genetic contribution than that for the number of eggs per year. Thus selection for egg weight would result in a faster change in total egg mass than selection for number of eggs per year. In other words, environmental conditions influence the number of eggs laid more than they influence the weight of eggs. Therefore, artificial selection would produce chickens that produce larger eggs, whereas controlling the environmental conditions such as lighting and feed would be more effective in increasing the number of eggs.

Twin Studies Are Helpful in Studying Polygenic Traits in Humans

For obvious reasons, heritability in humans cannot be calculated by making the environment constant ($V_E = 0$) for a large number of individuals. However, one way to determine the relative contributions of genes and environment to polygenic traits in humans is to study genetically identical individuals produced by nature: identical twins. Identical twins are produced when a single sperm fertilizes a single egg, and the fertilized egg (called a zygote), rather than dividing to form one individual, early in its development splits to form two individuals. Since these embryos are the products of the same fertilized egg, they are genetically identical. However, twins also share a similar environment, except for those rare identical twins that have been raised apart. Therefore, in studies of twins, the percentages of twins that share the same trait, called percent concordance, are compared between identical (or monozygotic) twins and nonidentical (or fraternal or dizygotic) twins. The dizygotic twins share a similar environment but on average share only half their genes. By accounting for the similar environments and the proportion of shared genes, twin studies provide an estimate of heritability. Table 12.1 shows examples of traits studied in mono- and dizygotic twins. The data are expressed as percent concordance, that is percent of twins in which both twins had same or similar measure of that trait.

Some behavioral and other traits that are not strictly physical are also polygenic. For example, propensity for diseases, longevity, and intelligence are polygenic traits. A large twin study from Sweden tested over one hundred pairs each of identical and fraternal twins over the age of eighty. Percent concordance of some of their measurements of

Table 12.1 Percent Concordance Among Twins

Trait	Monozygotic	Dizygotic
Height	93	64
Weight	92	63
Birthweight	67	58
Childhood asthma	65	40
Coronary artery disease	46	12
Schizophrenia	60	10

cognitive ability is shown in figure 12.4. Intelligence, like other behavioral or mental traits, is a complex trait that can be measured in different ways. As shown in the graph, different measures of cognitive ability exhibit different degrees of genetic contribution. Thus mental speed has a much higher heritability than spatial ability. For complex behavioral and disease traits, twin studies provide an important tool to determine the genetic contribution to human traits.

Quantitative Traits in Medicine and Agriculture

We saw in chapter 9 that susceptibility to disease is likely to be determined by defects in many genes. Autism seems to belong to that category of diseases. Autism is a neurodevelopmental disorder first recognized as a disease in 1943. It is already apparent during the first three years of life, expressed in lack of verbal communication, social responsiveness, and ritualized behavior. Autism affects approximately 1 child in 2,500.

Over the years, several theories have been put forth to explain this disease. These include behavioral, environmental, dietary, viral, autoimmune, and genetic causes. Statistical studies have shown that occurrence of autism among siblings is 2–6 percent, a value much higher than simple chance alone, which is 1 in 2,500, or, 0.004 percent. In addition, monozygotic twins show a twenty-five-fold higher concordance for autism than dizygotic twins. These results strongly suggest that autism is a polygenic trait.

An American-Australian team recently tackled the problem, this time using a total of 519 DNA markers to study 147 pairs of autistic siblings as well as their parents. They did a total of 160,000 DNA-typing experiments. The conclusions from their study are that fifteen or more genes seem to be involved in autism. These genes were

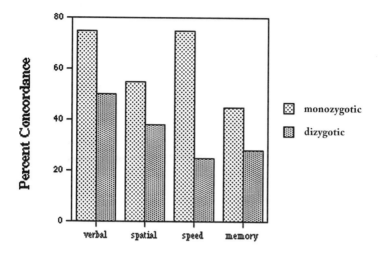

Cognitive Ability

Figure 12.4 Percentage of Twins with Similar Measures of Cognitive Ability. The data are from a Swedish study of twins at least eighty years old. As expected, identical, or monozygotic, twins share a higher percentage of all cognitive abilities than do fraternal, or dizygotic, twins. Similar to other polygenic traits, cognitive ability is complex, and some measures, like spatial ability, show less heritability than speed, for example.

mapped on chromosome 1, with a possible intervention of genes located on chromosome 17. They have a rough estimate of the number of genes involved and the rough location of a few of these genes on chromosomes. Because the human genome is huge and we can not do experimental crosses of humans, it may be years before the genes for autism are identified. However, the completion of the Human Genome Project will help to identify this and other genes involved in polygenic traits in humans.

To further illustrate the complexity of diseases influenced by many genes, it is believed that alcoholism, schizophrenia, and manic-depressive disorder are also at least partially controlled by genes. However, in spite of various claims, we still do not know how many genes are involved or where these genes are located in the human genome.

The situation is slightly different with agricultural plants and animals. Another name given to genes involved in multigenic traits is "quantitative trait loci," or QTLs. QTLs are responsible for phenotypes in plants, such as fruit weight, acidity, content of soluble solids, and so forth. Many QTLs have been mapped and some even cloned. The main reason for this success is that, unlike with humans, controlled crosses can be done with plants. Crosses tell us what kinds of phenotypes are obtained in the offspring, and the genes corresponding to these phenotypes can be mapped by using DNA markers and linkage analysis as discussed in chapter 9. Then individuals originating from various crosses are analyzed for polymorphisms at the DNA level, also as discussed in chapter 9. To use the example of the tomato, more than three hundred polymorphic loci have been identified in this plant, with six affecting fruit weight and five affecting acidity. Agronomists now conduct this type of research in a wide variety of edible plants in order to find the genes involved in many polygenic traits that affect the quality of our food.

Summary

Many important traits are determined by more than one gene. These traits are called additive because the contributions from many genes are added to produce the phenotype. These traits, also called multigenic, polygenic, or quantitative, can be recognized as those traits with continuous variation in phenotype. The inheritance of polygenic traits is complicated by the fact that many genes and environmental conditions contribute to the trait. For agricultural breeding, it is important to calculate the heritability, or the amount of genetic contribution, to multigenic traits. In humans, twins studies are helping to determine the genetic and environmental contribution to important human traits.

CHAPTER 13

Genetically Modified Animals and the Applications of Gene Technology for Humans

GENETICALLY MODIFIED PLANTS are now widely available to consumers, as discussed in chapter 6, and, as we have also seen, generate a significant amount of controversy. Yet genetically modifying plants may seem to many people to be much less controversial than tampering with the genetic blueprint of animals. This is due in part to our own view of animals. Indeed, for most people it is easier to relate to a dog, a goat, or even a mouse than it is to relate to a banana or a corn plant. What's more, the genetic engineering of animals immediately brings to mind the prospect of genetically modifying human beings. In this chapter we will see how scientists have created genetically modified animals and how gene cloning is being explored to treat genetic diseases in humans.

Cloning Animals by the Nuclear-Transfer Technique

Today, most genetically modified animals are produced by the nuclear-transfer technique, a method that produced Dolly the Scottish sheep, the very first cloned mammal. To date, all sorts of animals have been cloned this way, including mice, pigs, goats, rabbits, cats, and cattle. In this procedure, an egg is removed from a female animal, and its nucleus, which contains the chromosomes, is carefully sucked out with a very fine glass needle. This operation removes the original DNA from the egg. Next, a diploid cell (in the case of Dolly, this was an udder cell) isolated from another animal is fused with that egg using an electric shock. This procedure forces the chromosomes of the diploid cell to penetrate the egg. Since the egg was rid of its own (haploid) set of chromosomes, the result is an egg cell that now contains two sets of chromosomes, both sets originating from the donor fused cell. This egg can now be coaxed to behave like a fertilized egg. However, contrary to a regular fertilized egg, in this case, *the two sets of chromosomes come from only one parent,* the donor of the fused cell. Since the processes of removing the nucleus and fusion with a diploid cell may injure the egg, scientists normally manipulate several eggs in a single experiment, not just one.

Next, the manipulated eggs are allowed to divide a few times in a petri dish and are subsequently implanted into yet another female that, if all goes well, will deliver a litter of cloned animals, all copies of the individual that donated its chromosomes. The procedure is depicted in figure 13.1.

It should be noted that nuclear transfer in mammals still needs improvement to become a routine procedure. In addition, many live offspring originating from the cloning procedure are crippled or deformed in one way or another. Stillbirths are also frequent. The reasons for this are poorly understood, and cloning (including genetic modification) by nuclear transfer is far from being well established.

The above technique is used for cloning, but up to this point, no genetic modifications of the cloned animals have really taken place. The twist however, is that the cells used as chromosome donors in a cloning experiment *can* be genetically manipulated before fusion with an egg. In that case, udder cells, as in the example of sheep, can be modified in the test tube with a gene from another species. Here also, an electric shock is used to prompt cells to pick up the DNA presented to them in their growth medium. For example, the Scottish scientists who created Dolly produced other cloned sheep using udder cells modified by the addition of gene for human blood-clotting factors. These cloned sheep did indeed produce the factors in their milk.

Genetically Modifying Animals Using Embryonic Stem Cells

The drawback mentioned for cloning animals by nuclear transfer does not exist for a second technique used to produce genetically

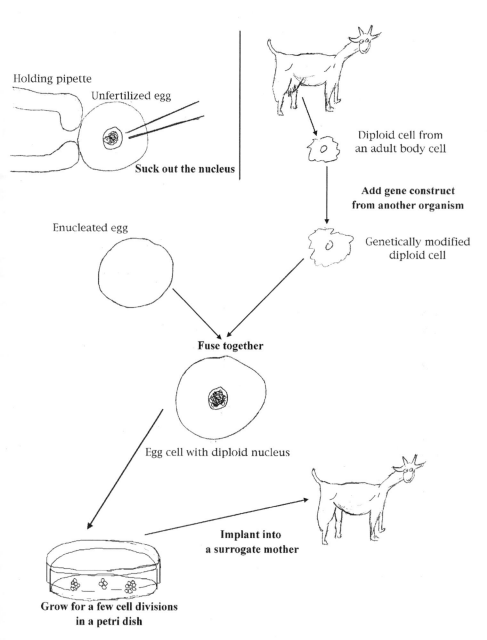

Figure 13.1 Making Genetically Modified Animals by Nuclear Transfer. First, the haploid nucleus of an unfertilized egg is sucked out of the egg with a very fine glass syringe. The enucleated egg is then fused with a diploid cell from the body of an adult animal. The egg, containing a diploid nucleus, is stimulated to divide in a petri dish. After a few divisions, the growing embryo is implanted into a surrogate mother that will carry it to term. The resulting cloned animal can be genetically modified by inserting a foreign gene into the diploid cell prior to fusion with the enucleated egg.

modified animals. This is a technique that genetically manipulates embryonic stem cells in the test tube and then injects them into a growing embryo. We saw in chapter 4 that stem cells isolated from very young embryos have the ability to form any kind of tissue. When injected into an embryo at the blastocyst stage (an embryo containing dozens of cells that have not yet formed any tissues or organs), embryonic stem cells join the cells of the developing embryo. Later, they follow the developmental and differentiation pattern of the cells surrounding them. In other words, an injected embryonic stem cell that finds itself next to cells destined to become a heart will also develop into a heart cell. If the embryonic stem cell finds itself next to cells destined to become gametes in the adult, it too will become a gamete.

Embryonic stem cells can be cultivated in the lab and manipulated like any other kind of cells. Thus, scientists can genetically modify embryonic stem cells by adding cloned genes from any source (usually by subjecting them to an electric shock in the presence of DNA) and then inject these genetically modified embryonic stem cells into a blastocyst-stage embryo. The blastocyst is subsequently implanted into a surrogate mother, where it continues to develop. Genetically modified embryonic stem cells thus become an integral part of the growing embryo and, eventually, the adult individual. These adults are then formed partly from the cells of the embryo that was the blastocyst and partly from the injected stem cells, and thus they are not all genetically alike. Thus, if the embryonic stem cells were genetically modified, only the cells derived from those embryonic stem cells have the foreign gene. The full procedure is illustrated in figure 13.2. It should be emphasized that genetically modified animals are not yet used on an industrial scale and their products are not yet used for medical purposes.

Uses of Genetically Modified Animals

The basic idea in making genetically modified animals is to turn animals into specialized protein factories, much like what was described for bacteria in chapter 5. The product made by this technique that may be furthest along in development is a blood-clotting factor produced in milk. However, some companies are trying to genetically engineer animals to produce useful proteins in their urine. This may seem repulsive, but there are some real advantages in producing compounds that are secreted in urine. While only females

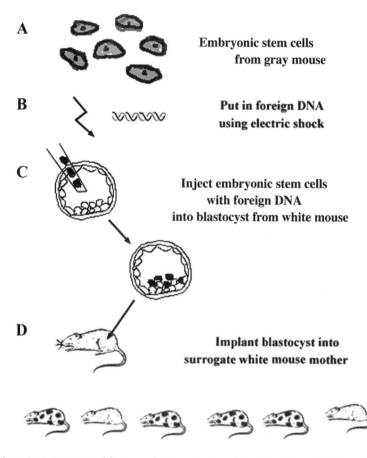

A Embryonic stem cells
from gray mouse

B Put in foreign DNA
using electric shock

C Inject embryonic stem cells
with foreign DNA
into blastocyst from white mouse

D Implant blastocyst into
surrogate white mouse mother

Figure 13.2 Genetic Modifications of Embryonic Stem Cells Using Mice of Different Colors. Note that the drawings are not to scale, the DNA, of course being much smaller than the cell, and the blastocysts being much smaller than the mice. A. Embryonic stem cells are obtained from the embryo of a gray-colored mouse. these cells are not actually gray; only the coat of the mouse is gray. The cells are shaded so that they can be distinguished from those of the white mouse. B. A foreign gene is transferred to the embryonic stem cells by electric shock. C. Embryonic stem cells with the foreign gene are injected into an early-stage embryo called a blastocyst. In this example, the blastocysts are derived from a white mouse so that we can distinguish the cells that have the foreign gene from those that do not. D. The blastocysts are then implanted into a surrogate mother that produces offspring. Some of the offspring are made up of a mixture of white-mice cells of the blastocyst and gray-mouse cells with the foreign gene, and for this reason they are called mosaics. If the cells from the gray mouse make the gametes, the sperm or the egg in these mosaics, breeding them will produce mice with the foreign gene in all their cells.

produce milk, both males and females produce urine, and urine is much less complex chemically than milk, so it may be easier to purify proteins produced in urine instead of milk. So far, experiments have produced genetically modified mice that produce small amounts of human-growth hormone. In addition to medically important proteins, scientists have also engineered animals that secrete other proteins of commercial interest, such as cows that produce silk protein in their milk.

One can imagine that animals could be engineered for innumerable human medical purposes. In fact, pigs have been engineered to produce human hemoglobin for blood transfusions. Other pigs have been genetically modified so as *not* to express antigens recognized as foreign by humans. The organs of these engineered pigs could potentially be used as transplants in human patients because they would not be rejected. However, so far such applications have not yet been implemented. One danger is that viruses that infect animals could easily and inadvertently enter humans through the organs or blood products from engineered animals. Such viruses may not be transmitted easily now because of natural barriers to transmission, but organ transplant will facilitate transmission. Because we know very little about pig viruses, we do not know the risks of a pig virus causing a major disease in humans. If even a single pig virus caused disease in humans and was easily transmissible, we could be faced with a catastrophic epidemic. Scientists are well aware of these risks and take them very seriously. In short, animal genetic modification is still at an experimental stage. At this point, it is unclear whether genetically modified animals should be or will ever be used as tissue donors, protein factories, or for other purposes.

Human Gene Therapy

It may seem undignified to discuss humans in a chapter devoted to genetically modified animals, yet from a purely biological viewpoint, humans are no different than other mammals in regard to science's ability to modify their genetic material. As you know, it is possible to fertilize human eggs with sperm in a laboratory dish and implant the young, dividing embryo into the uterus of the future mother. That mother may or may not be genetically related to the embryo. This technique is called in vitro fertilization, or IVF. This technique is now routinely used to help infertile couples have children.

As we saw, some animals have been engineered with foreign genes, one such gene being the one that codes for a blood-clotting factor. Why not then, one can reason, directly add a functional blood-clotting factor gene to a fertilized human egg that would otherwise become a hemophiliac? At this point, one answer is that this technique would be too dangerous. A point not mentioned previously is that when we insert a foreign gene into an organism, this gene randomly inserts into the genome of the host cell. It does not necessarily replace a copy of that gene in the genome, defective or not. A foreign gene can, by chance, insert into the middle of a gene, effectively mutating it. Thus, the blood-clotting factor gene injected into an egg could insert itself in a place of the genome where it could disrupt other genes. The effect of this disruption would be totally unpredictable; we would not know what gene might be affected and what the effect of the disruption would be on the organism. Another objection that many have is that changing the genetic features of a whole individual will result in genetically altered future generations of this individual. This change in the gametic cells, as compared to change in all the other cells of our body that will not produce future generations, is considered taboo by many.

Human beings can be genetically altered by adding genes to *some of the cells* of an affected individual rather than manipulating the egg, a procedure that will alter the genes of all the cells in an individual, including the cells destined to be gametes. Thus, gene therapy is being pursued for diseases in which a *single gene defect affects primarily one type of tissue whose cells are accessible.* In order for the disease to be treatable by this method, we first need to understand which gene is defective. This is an arduous process that we briefly discussed in chapter 9. Once the genetic defect is identified and the normal copy of that gene is isolated, there are two different approaches toward gene therapy. One method is to remove some of the affected cells or cells that will develop into affected cells from the patient, genetically modify those cells, and then return them to the patient. This type of procedure is called ex vivo gene therapy. The second method is to treat the patient directly through DNA injection into his or her organs. This type of procedure is called in vivo gene therapy.

Ex vivo gene therapy is used for diseases that arise from problems with blood or bone marrow cells. These cells, such as our red blood cells and cells of the immune system, circulate in our blood stream.

The very first case of human gene therapy used this ex vivo approach. It was applied to a genetic condition called severe combined immunodeficiency (SCID). This disease results from a mutation in a gene coding for the enzyme adenosine deaminase (ADA). In the absence of ADA, affected individuals cannot develop a functioning immune system and soon die from even minor infections. The first SCID patient treated with gene therapy was a young girl named Ashanti DiSilva. Some of Ashanti DiSilva's T cells (important components of the immune system) were removed from her blood, treated with a correct ADA gene, and reinjected into her body. Today, Ms. DiSilva leads a normal life, over twelve years after her treatment, and her ADA levels are normal. Since this pioneering clinical trial, several additional children with this condition have been treated with gene therapy. However, recently three of the treated children developed the same unusual form of leukemia, a cancer of the blood cells. It seems that all three diseases were caused by the insertion of the corrective gene into or near a cancer-causing gene. This is one danger of gene therapy, and a danger that is unpredictable. Thus gene therapy is still at an experimental stage even for cells that can be easily removed and added back to the patient.

A more general approach to gene therapy is to give a dose of the correct gene directly to the patient. However, this requires a safe vector that can target the gene to the right cells. Recall from chapter 5 that vectors are pieces of DNA that carry foreign DNA into cells. In gene therapy these vectors can be either plasmid or viral and must be able to carry the good copy of the gene to the cells that are affected. Viruses were considered good candidates for vectors for human gene therapy because some viruses, by their very nature, infect only a certain type of cells. For example, the flu virus typically infects cells in our respiratory tract and lungs. Yet an important consideration is that no negative effects be caused by the procedure. This is especially important if the vectors are viral DNA. Indeed, viral vectors must be able to invade the target cells, but they should not cause disease or an allergic response. Because of the potential dangers of viral vectors, some researchers are testing other type of vectors that are not of viral origin to transport DNA into cells.

Viral DNA vectors have already been used in human gene therapy trials aimed at correcting metabolic disorders. One such vector is DNA isolated from adenovirus, a rather benign virus responsible for one form of the common cold. When injected into the liver, for example, this virus is not expected to cause any symptoms. Unfortunately, this is not what happened to Jeff Gelsinger, a patient suffering from a defect in the gene that codes for an enzyme called ornithine transcarbamylase. This enzyme removes ammonia from the blood. The correct version of the gene was cloned into an adenovirus vector and injected into Jeff Gelsinger's main liver artery. Shortly after the treatment, he started showing severe allergic response and died soon after. The autopsy showed that the viral DNA had propagated to all his organs and had triggered a massive immune response that killed him. As a result, clinical trials of human gene therapy using an adenovirus vector have been stopped. However, research continues with other viral vectors, particularly for the treatment of cystic fibrosis, a disease that has major manifestations in the lung cells. Clearly, much work is needed to ensure that gene therapy can be done safely.

Nevertheless, other gene-therapy trials are currently in progress. In the case of hemophilia, for example, one could insert a human blood-clotting factor gene into a viral DNA vector and inject the recombinant virus into the liver of the patient. If this procedure succeeded, the liver cells would make enough of the factor to ensure correct clotting of the blood. The first clinical trials of gene therapy for hemophilia were done on patients who made less than 1 percent of the normal amount of clotting factor. Though initial work was done more to test the safety of the procedure than to see if the disease could be cured, the three patients began to produce normal clotting factor. Though not cured, their symptoms were less severe.

Heart specialists also use gene therapy on a limited scale. It has been shown that the gene coding for vascular endothelial growth factor (VEGF), when injected into human muscle cells promotes the formation of blood vessels. As we know, many heart conditions are due to plugged-up arteries that can no longer supply the heart with oxygen. Injection of the VEGF gene into the heart can alleviate pathological symptoms through the formation of new blood vessels in the heart itself.

Human Reproductive Cloning

Stimulated by the success of animal cloning by nuclear transfer, some scientists have recently announced their intention to clone human beings by using the same technique. Basically, they would isolate nuclei from the cells of the donor wishing to be cloned, inject a diploid nucleus into an enucleated egg obtained from a volunteer female, and implant this egg into the womb of a surrogate mother. If all went well, the baby would be an almost exact genetic copy of the donor of the nucleus. We say "almost exact" rather than "exact" because human beings possess two types of genomes. The main genome resides in the cell nucleus and contains about 3.15 billion base pairs. However, human cells also contain mitochondrial DNA. Mitochondria are responsible for the production of energy and can be considered the energy factories of the cell. These bodies harbor a short piece of DNA that contains only about 16,600 base pairs. Thus, a human clone would still contain genes from mitochondrial DNA originating from the mitochondria present in the enucleated donor egg.

In principle, there is no reason to believe that human cloning would not work, given that it does work in several other mammalian species. However, it is clear that human cloning raises major ethical questions, simply because human beings and human embryos are involved. First, there is the high risk of fetal malformation as seen in the cases of all presently cloned animal species. Do we have the right to produce individuals that may end up crippled because of the procedure that led to their creation? Then there is the potential psychological impact on a clone (and perhaps on the nuclear donor) of being the almost perfect genetic copy of one of his or her parents, while sharing little with the egg donor. Further, there are the horrendous complications regarding the kinship of a clone. For example, let us assume that the nuclear donor is a man. His cloned boy will be his son, but this son will have no mother, only a paternal grandmother in the female category. And how should the clone look at the woman who donated the enucleated egg? Would she be his mother? In a genetic sense, this "mother" would have only contributed mitochondrial DNA. Finally, since human cloning is reputedly sought mostly by infertile couples, the birth mother of the clone could very well be a surrogate, totally genetically unrelated to the "parents." In other words, the clone would have two "mothers" (the enucleated egg donor and the surrogate that brought him to term) and one father of whom he is a copy. Many other situations involving cloning can be imagined, and you are invited to do so.

Human Therapeutic Cloning

All the mostly negative press given to human reproductive cloning has unfortunately obscured the potential benefits of cloning in the medical arena. The technique we will describe below does not directly involve human reproduction, but it uses human embryos as cell donors for medical purposes. Since human embryos are destroyed in the process, this technique is not without controversy either.

We saw in this chapter that genetic diseases could, in principle, be remedied by injection of a cloned "good" gene. However, not all diseases are amenable to treatment by gene injection. For example, heart disease, degenerative kidney disease, and arthritis (also a degenerative disease), as well as spinal-cord injuries, are not the result of the malfunctioning of a single gene. Going back to embryonic stem cells (see chapter 4), we have seen that they can differentiate into any kind of specialized tissue. Given that, they might be able to regenerate damaged heart, spinal-cord, kidney and joint tissues if injected into these organs. This has indeed been shown to a certain extent in other animals, such as the mouse. Why not then apply this technique to humans? The first problem is that embryonic stem cells from a donor will not be immunologically compatible with the cells of the recipient. These cells would quickly be rejected, though there is a way around this. Embryonic stem cells could be derived from an embryo that was the result of a nuclear transfer from the person suffering from the disease. In this case, since the embryo would be practically genetically identical to the nuclear donor, rejection would not happen.

However, a major problem is that one must first create this human embryo and then dissect it in order to use its cells. This procedure is not compatible with embryo survival. For some, this practice is equivalent to murder and cannot be condoned. For now, stem-cell research from cloned human embryos cannot be conducted with federal funds in the United States. Nevertheless, a private U.S. company has announced partial success in human embryo cloning by nuclear transfer, and research is also proceeding in other countries. The future of cloned embryonic stem cells in human medicine is presently uncertain.

Summary

The new genetic technology carries enormous potential for medical applications. These applications range from utilizing animals as biofactories to human gene therapy and to human therapeutic cloning. All the basic techniques to achieve these goals are available but must be refined. The use of human embryos, either for reproductive cloning or for therapeutic cloning, is highly controversial, however. As is often the case with new technologies, the good often comes with the bad. We hope that a thorough understanding of the genetic principles explained in this book will allow you to make objective decisions regarding all the potential—and sometimes unsettling—applications of genetics.

Internet Resources

Web Sites Applicable to Many Topics

Cold Spring Harbor Laboratory Dolan DNA Learning Center: A great site with information about DNA and genetic diseases
http://vector.cshl.org/

A particularly good set of animations are found at this site at
http://vector.cshl.org/resources/BiologyAnimationLibrary.htm

Virtual Library of Genetics sponsored by the Department of Energy
http://www.ornl.gov/TechResources/Human_Genome/genetics.html

Chapter 2

Electronic Scholarly Publishing: This growing resource has one of the best collections of information on the foundation of genetics. There are original papers in genetics, a book on the history of genetics, and a timeline of genetic discoveries put in the context of other historical events.
http://www.esp.org/

MendelWeb: An excellent site featuring the original papers of Gregor Mendel and English translations with abundant annotations, explanations of experimental methods, and statistical analyses.
http://www.netspace.org/MendelWeb/

Biology Labs On-Line Free Trial site: This site includes FlyLab, which allows you to cross different mutants of fruitflies
http://www.biologylab.awlonline.com/trial.html

Interactive Punnett Square
http://www.athro.com/evo/gen/punexam.html

Chapter 3
Online Mendelian Inheritance in Man: An official database of human genes and genetic disorders.
http://www.ncbi.nlm.nih.gov/Omim/

Gene Clinics: A publicly funded information resource on genetic diseases.
http://www.geneclinics.org/

Center for Disease Control: Official U.S. government site, with information about many diseases and statistics.
http://www.cdc.gov

Many organizations have Web pages to help individuals affected with genetic diseases:

The National Institutes of Health has an information page on hemochromatosis.
http://www.niddk.nih.gov/health/digest/pubs/hemochrom/hemochromatosis.htm

The American Hemochromatosis Society also maintains a Web page.
http://www.americanhs.org/

Lineages Web site.
http://www.lineages.com

Many states have Web sites for their newborn-testing programs. Examples include:

Oregon (also covers Idaho, Nevada, Alaska, Delaware, Hawaii, and some military facilities).
http://www.ohd.hr.state.or.us/nbs/

Wisconsin.
http://www.slh.wisc.edu/newborn/

Chapter 4
National Center for Biotechnology Information.
http://www.ncbi.nlm.nih.gov/

Chapter 5

BIOPHARMA—a resource on pharmaceutical products including those made via recombinant DNA technology.
http://www.biopharma.com/book_contents.html

Chapter 6

Food and Drug Administration.
http://www.fda.gov

Consultative Group on International Agricultural Research.
http://www.cgiar.org

International Rice Research Institute.
http://www.irri.org

Chapter 8

Occupational Safety and Health Administration.
http://www.osha.gov

Chapter 11

U.S. National Germplasm resource Web page.
http://www.ars-grin.gov/

Glossary of Scientific Names of Organisms

Scientific names for organisms are italicized and have two parts. The first is the genus (plural genera) name, which is capitalized. Genus refers to a group of organisms that are closely related. The second part is the species name.

Achillea millefolium Milfoil or yarrow. Aromatic perennial plant in the composite family with finely divided leaves.

Agrobacterium tumefaciens Soil bacterium that causes tumor formation in plants called crown gall disease. This bacterium has a plasmid responsible for tumor formation called Ti plasmid, which is widely used for genetic engineering in plants.

Arabidopsis thaliana Small weedy plant in the mustard family. It has become a model plant for genetics since it grows easily and has a relatively small genome, which has been sequenced.

Bacillus thuringiensis Bacterium that is a natural pathogen of insects. It produces toxins that kill insects' intestinal cells.

Caenorhabditis elegans Tiny (˜ one mm in length) nematode that has become a model animal for genetic and developmental studies. Its developmental pattern has been determined and its genome has been sequenced.

Drosophila melanogaster The fruit fly that been used for many different genetic studies. Its genome has been sequenced.

Erwinia ureidovora Soil bacterium that is a source of one of the genes used to make golden rice, which produces provitamin A in the seed.

Escherichia coli Common intestinal bacterium.

Fugu rubripes Pufferfish; considered a model genetic animal for vertebrates.

Homo sapiens neanderthalanis Neanderthals.

Homo sapiens sapiens Modern humans.

Streptococcus pneumoniae Bacterium that causes pneumonia.

Thermus aquaticus Bacterium from which heat-resistant DNA polymerase was isolated. This DNA polymerase is used in polymerase chain reaction.

Glossary of Human Genetic Diseases

Below are brief descriptions of the human genetic diseases mentioned in this book. More information about these and other human genetic diseases can be obtained from http://www.ncbi.nlm.nih.gov/Omim

Amish diseases A variety of very rare diseases are found among the Amish due to their small population size and reproductive isolation. *See* nemaline myopathy and brittle hair syndrome.

Alkaptonuria A recessive trait characterized by urine that turns dark upon exposure to air.

Brittle-hair syndrome One of the Amish diseases; also called hair-brain syndrome. Characterized by short stature, mental deficiency, brittle hair, and reduced fertility.

Cancer A generic term to describe uncontrolled cellular growth. BRC1 and BRC2 are two genes that greatly increase the chance for breast cancer. HPC is a gene that, if defective, greatly increases the chance for prostate cancer

Color blindness There are several forms of color blindness. The most common X-linked form results in red/green color blindness.

Diabetes A group of diseases that cause malfunctions in dealing with sugar in the blood stream. Type I, or diabetes mellitus, is the insulin-dependent form. This disease has contributions from both genetics and environment. Type II is an adult-onset, non-insulin-dependent form caused by the actions of several genes.

Fragile X An X-linked dominant trait characterized by mental deficiency. Called fragile X because of a constricted region on the X chromosome, called the fragile site, in the karyotypes of affected individuals observed under some conditions. Due to triplet repeat expansion.

Galactosemia A disease of the ability to digest galactose, which is one of two sugars making up the milk sugar, lactose. Babies begin vomiting and having diarrhea within days of drinking milk and often die early. If detected early, individuals can live relatively normal lives by having a strict lactose- and galactose-free diet.

Hemochromatosis A common disease caused by excess iron in the body that manifests itself in a variety of ways including joint pain, diabetes, cirrhosis of the liver, liver cancer, and heart failure. The recommended treatment is bloodletting.

Hemophilia A genetic disease characterized by the blood's inablity to clot. There are several forms of this disease, including hemophilia A and B, which are X linked.

Huntington's disease A dominant degenerative nerve disease with adult onset. It is also called Huntington's chorea due to the choreic or shaky movements of affected individuals.

Maple syrup urine disease Also called fenugreek urine disease because of the characteristic odor of the urine.

Marfan syndrome A dominant disease of the connective tissues. Individuals are characterized by a tall lanky stature and long digits. The main medical problem is due to a weakened aorta that can rupture without warning.

Muscular dystrophy Also called myotonic dystrophy; characterized by muscle weakness and wasting. A dominant trait caused by expansion of triplet repeats.

Nemaline myopathy One of the Amish diseases; a neuromuscular disease characterized by muscle weakness and "nemaline" or threadline structures in the skeletal muscles.

Phenylketonuria Abbreviated PKU; a recessive trait causing an inability to metabolize the amino acid phenylalanine.

Sickle-cell anemia A recessive trait characterized by sickle-shaped red blood cells caused by defective hemoglobin.

Tay-Sachs disease A recessive neurodegenerative disease that is fatal by two to three years of age.

Glossary of Terms

Amino acid Building block of protein molecules. Also important components of our diet.

Amniotic fluid The liquid surrounding a fetus in its mother's womb; enclosed by the amniotic membrane. The amniotic membrane and its contents constitute the amnion.

Antibiotic A chemical that inhibits bacterial or fungal proliferation.

Antibody A protein molecule synthesized in the cells of the immune system whose role it is to neutralize antigens.

Anticodon The portion of a tRNA molecule that recognizes a specific codon.

Antigen Any substance recognized as foreign by the immune system.

Bacteriophage Also called a phage. A virus that infects bacteria.

Base substitution A change of base pair in DNA.

Biolistics A technique to introduce DNA into plant cells. Metal particles on which DNA is adsorbed are shot into cells, usually by means of a gas-driven piston.

Blastocyst A young embryo composed of several dozen cells containing a cavity in its center.

Catalyst A compound that accelerates chemical reactions.

Chorionic villus Embryonic tissue surrounding the amnion.

Chromosome A structure composed of DNA and protein which is visible under the microscope in dividing cells.

Clone Identical copies of DNA, or genetically identical cells or organisms.

Cloning 1. The act of isolating and multiplying a certain piece of DNA containing one or more specific genes. 2. The act of producing one or several identical cells or organisms from a single cell or organism.

Codominance A situation in which two different versions of the same gene are equally expressed and have equal contribution to phenotype.

Codon A set of three contiguous base pairs in DNA (or three bases in RNA) that corresponds to a specific amino acid. There are sixty-four possible codons from the four DNA building blocks arranged in sets of three.

Complementary base pairing Pairing between A and T or G and C in DNA or A and U or G and C in RNA.

Cross A sexual mating producing offspring.

Deletion A mutation caused by the removal of one or more base pairs in a gene.

Deoxyribose A six-carbon sugar present in DNA.

Diploid A cell or organism containing two complete sets of chromosomes.

DNA ligase An enzyme used to glue together two pieces of DNA.

DNA polymerase The enzyme that replicates DNA.

Dominant A gene or trait that is phenotypically expressed and masks the phenotype of another gene or trait.

Drift In small populations, refers to the fact that by chance alone some gene variants can become prevalent while others can become rare or go extinct.

Electrophoresis A technique that separates molecules such as proteins and DNA fragments in an electric field.

Embryo An organism containing tens of cells resulting from the fusion of an egg and a sperm cell. An embryo develops into a fetus after many cell divisions.

Enzyme A biological catalyst that accelerates a specific chemical reaction.

Evolution Modification of organisms by a selection process acting on variation among individuals.

DNA fingerprinting A technique that can identify a single individual by his/her/its DNA profile.

Fitness The ability of individuals with particular phenotypes to pass on their genes to the next generation. This ability is dependent on both the genetic makeup of the individual and the environment.

Founder effect The genetic results of the creation of a new population by a few members from a larger population.

Gametes Haploid cells that fuse to form an individual. Sperm, eggs, pollen grains, and ovules are gametes.

Gene The basic unit of heredity. Genes are composed of DNA.

Gene therapy A method of treating a genetic disease by providing a patient with a correct version of a defective gene.

Genetic code The code containing the information in DNA used to produce proteins.

Genetic engineering The act of introducing one or several new genes into an organism.

Genetically modified organism Any organism containing one or several genes normally not found in that organism. The introduced gene can be from another organism or a gene from the same organism with modifications.

Genome The suite of all genes in an organism.

Genotype The types of gene combinations, dominant or recessive, of an organism. Often shown for a single trait or a few traits using upper- and lowercase letters.

GMO *See* genetically modified organism.

Haploid A cell or organism containing only a single set of chromosomes.

Hermaphrodite An organism in which one individual produces both male and female gametes. Some can self-fertilize. Most plants and some animals are hermaphrodites.

Heterogametic Individuals that have two different types of sex chromosomes. In humans, the heterogametic sex is male, with an X chromosome and the Y chromosome. Among birds, the heterogametic sex is female, with two different sex chromosomes.

Heterozygote Individual containing two different copies of a given gene, such as **Aa**.

Homogametic Individuals that have two of the same type of sex chromosomes. In humans, the homogametic sex is female, with two X chromosomes. Among birds, the homogametic sex is male with two matching sex chromosomes.

Homozygote Individual containing two identical copies of a given gene, such as **AA** or **aa**.

Insertion A mutation caused by the addition of one or more base pairs to a gene.

Linkage A situation in which genes are located on the same chromosome.

Meiosis The special cell division that produces gametes. The process produces four haploid cells, that is, cells with half the number of chromosomes of the diploid progenitor cell.

Messenger RNA Also abbreviated mRNA; an RNA copy of a DNA gene made by the process of transcription.

Metabolic disease A disease in which affected individuals are unable to metabolize certain compounds present in the cells.

Migration Population movement that results in mixing one population with another.

Mitochondrion (pl. mitochondria) Cellular organelles that generate energy using oxygen.

Mitosis Cell division that produces two cells identical in genetic composition to the progenitor cell.

mRNA *See* messenger RNA

Mutagen A substance or physical effect that causes mutations above and beyond the spontaneous mutation rate.

Mutation A change in DNA base-pair sequence.

Natural selection Conditions in nature that increase or decrease the chance for individuals of certain phenotypes to pass on their genes to the next generation.

Nitrogenous base A term used to refer to the part of the DNA and RNA molecules that include A, G, C, T and U. They are called nitrogenous bases because they contain nitrogen atoms.

Nucleotide The combination of a nitrogenous base with ribose or deoxyribose and a phosphate group.

Nucleus The part of the cell where most of its DNA is located.

Palindrome A word or DNA sequence that reads the same forward and backward. With DNA, the sequence reads the same going forward in the one strand and backward in the other strand of the double helix.
Example: GCATATGC
         CGTATACG

Pathogen An organism that causes disease.

PCR *See* polymerase chain reaction.

Pedigree A compendium of relationships among individuals in a family; a family tree.

Phenotype The physical properties of an organism. Many phenotypes are under the control of genes, but some are not. Examples of phenotypes are: flower color in plants, the ability or inability to metabolize the sugar galactose in humans, coat color in dogs, and so on.

Phenylketonuria A genetic disease characterized by the inability to metabolize the amino acid phenylalanine.

Phytoremediation The action of cleaning polluted soils with plants.

Plasmid A circular piece of double-stranded DNA much smaller than the genetic material in chromosomes. Plasmids are used for genetic engineering.

Pluripotent A cell that can differentiate into many different types of cells in the body.

Polymer A substance made of a large number of repeating subunits.

Polymerase chain reaction Abbreviated PCR; a technique allowing DNA replication in the test tube.

Polymorphism Things that exist in different variants in the population, for example, genes or DNA sequences.

Population A collection of individuals found in the same geographic location and where mating among individuals is possible.

Primer A short piece of single stranded DNA complementary to a piece of DNA that allows initiation of its replication.

Promoter A DNA sequence situated just before a gene whose function is to bind RNA polymerase.

Proofreading enzymes enzymes that scan newly synthesized DNA for potential base-pairing errors.

Protein A polymer composed of amino-acid building blocks.

Random genetic drift *See* drift.

Recessive The opposite of dominant. A recessive trait is only seen, that is, phenotypically expressed, in an individual without a dominant trait. This situation arises when one has two copies of the recessive trait or when the heterogametic sex has just one copy of the recessive trait on one sex chromosome.

Recombinant DNA DNA made in the test tube containing genes originating from different sources.

Recombination Shuffling genetic information by cutting and pasting DNA strands.

Replication The act of copying each strand of the DNA to produce two double-stranded DNA from one double-stranded DNA.

Restriction enzyme An enzyme that cuts double-stranded DNA at a specific sequence.

Ribose A six-carbon sugar found in building blocks of RNA.

Ribosome A small cellular structure that makes proteins encoded by mRNA and is thus involved in translation. The ribosome is composed of RNAs and proteins.

RNA polymerase An enzyme that makes RNA. It can make RNA copies of DNA genes through the mechanism of transcription.

Segregation Separation of two copies of a gene in meiosis.

Selfing The action of self-fertilization. Can commonly occur in the plant world.

Strain A collection of identical cells derived from one single cell and possessing the same phenotype and genotype. A strain is in fact a clone.

Template A DNA strand that serves as blueprint during DNA replication and transcription.

Teratogen A substance that produces malformations in a developing fetus without affecting its genetic material.

Terminator A DNA sequence at the end of a gene that indicates the end of the gene and signals the end of transcription.

Trait A particular phenotype such as eye color, height, hairy leaves, and so on.

Transcription The act of making an RNA copy of a DNA gene.

Transfer RNA A small RNA molecule that binds a specific amino acid and deciphers a codon in the mRNA.

Transformation The act of introducing DNA into living cells.

Translation The act of making protein molecules encoded by messenger RNA.

Triploid A cell or organism containing three sets of chromosomes.

Zygote The product of fertilization of an egg by a sperm cell.

Index